PROBLEMS IN PACKAGING:
The Environmental Issue

ELLIS HORWOOD SERIES IN ENERGY AND FUEL SCIENCE

Series Editors: I. BOUSTEAD and G. F. HANCOCK, Faculty of Technology, The Open University, Milton Keynes

HANDBOOK OF INDUSTRIAL ENERGY ANALYSIS
I. BOUSTEAD and G. F. HANCOCK, The Open University, Milton Keynes.

ENERGY AND PACKAGING
I. BOUSTEAD and G. F. HANCOCK, The Open University, Milton Keynes.

ENERGY AND MATERIALS EFFICIENCY OF RECYCLING PROCESSES
I. BOUSTEAD and G. F. HANCOCK, The Open University, Milton Keynes.

PROCESSING OF NORTH SEA OIL
P. LEWIS, The Open University, Milton Keynes.

LIQUEFIED PETROLEUM GASES
Guide to Properties, Applications and Uses: 2nd Edition – Revised and Extended
A. F. WILLIAMS, Principal Scientist, Esso Research Centre, Abingdon, and W. L. LOM, former Senior Scientific Associate, Esso Research Centre and Technical Adviser on Energy Matters to the European Investment Bank, Luxembourg.

WAVE ENERGY: A Design Challenge
R. SHAW, Senior Lecturer, Department of Mechancial Engineering, University of Liverpool

PROBLEMS IN PACKAGING: The Environmental Issue
I. BOUSTEAD, Faculty of Technology, The Open University, Milton Keynes, and K. LIDGREN, University of Lund, Sweden

PROBLEMS IN PACKAGING: The Environmental Issue

Editors:

I. BOUSTEAD, B.Sc., M.Sc., Ph.D.
The Open University, Milton Keynes

and

K. LIDGREN, Ph.D.
University of Lund, Sweden

ELLIS HORWOOD LIMITED
Publishers · Chichester

Halsted Press: a division of
JOHN WILEY & SONS
New York · Chichester · Brisbane · Toronto

First published in 1984 by

ELLIS HORWOOD LIMITED
Market Cross House, Cooper Street, Chichester, West Sussex, PO19 1EB, England

The publisher's colophon is reproduced from James Gillison's drawing of the ancient Market Cross, Chichester.

Distributors:

Australia, New Zealand, South-east Asia:
Jacaranda-Wiley Ltd., Jacaranda Press,
JOHN WILEY & SONS INC.,
G.P.O. Box 859, Brisbane, Queensland 40001, Australia

Canada:
JOHN WILEY & SONS CANADA LIMITED
22 Worcester Road, Rexdale, Ontario, Canada.

Europe, Africa:
JOHN WILEY & SONS LIMITED
Baffins Lane, Chichester, West Sussex, England.

North and South America and the rest of the world:
Halsted Press: a division of
JOHN WILEY & SONS
605 Third Avenue, New York, N.Y. 10016, U.S.A.

British Library Cataloguing in Publication Data
Problems in packaging. –
(Ellis Horwood series in energy and fuel science)
1. Packaging – Law and legislation – Europe
I. Boustead, I. II. Lidgren, Karl
342.4'42

Library of Congress Card No. 84-3743

ISBN 0-85312-721-2 (Ellis Horwood Limited)
ISBN 0-470-20068-5 (Halsted Press)

Typeset by Ellis Horwood Limited.
Printed in Great Britain by Unwin Brothers of Woking.

Table of Contents

Table of Contents 5

Preface 7

Chapter 1 Introduction. 9
I. Boustead

Chapter 2 The physical basis of restrictive legislation 11
I. Boustead

Chapter 3 A conceptual framework for a social appraisal of the beverage container issue 26
J. C. D. Fisher

Chapter 4 Measures to promote re-use or recycling of beverage containers . . 46
P. Lieben

Chapter 5 Packaging legislation in Denmark 60
K. H. Garmin

Chapter 6 Product charges in Norway – experiences from the packaging sector 65
H-C. Holdt

Chapter 7 An industrial view of the problems of public intervention in the field of packaging in Finland 72
J. Hämäläinen

Chapter 8 Present and impending regulations concerning packaging and the environment in the Netherlands 79
A. Th. van Ewijk

Chapter 9 Limitation of energy consumption and domestic waste production in the field of beverages 84
J-C. Cohadon

Chapter 10 Recycling of aluminium cans – from voluntary system to public intervention 110
M. Backman

Chapter 11 The Swedish container levy – an evaluation 124
K. Lidgren

Chapter 12 Evaluation of one-way cartons and returnable bottles for milk – economic aspects and ecological background 140
H. Wietbrauk

Chapter 13 The glass milk bottle is less healthy than cartons 153
W. J. M. Duyvestijn

Chapter 14 Evaluation of the approach to the problems of milk packaging and the environment in the Netherlands. 157
D van Doorn & H. Middelveldt

Index 172

Preface

The papers presented here in book form were read at an invitation conference in Sjöbo, Sweden organised by the TEM group of the University of Lund in June, 1982. In preparing them for publication, authors were asked to update any information relevant to their contributions and editing was limited to minor grammatical corrections and, in some cases, shortening of the papers. None of the essential arguments presented at the Conference have been omitted.

In preparing the final manuscript, I am indebted to my secretary, Mrs Pat Brittin, and to our publisher, Ellis Horwood Ltd, for their tolerance in coping with the innumerable small changes.

I. Boustead
1983

CHAPTER 1

Introduction

I. BOUSTEAD
Faculty of Technology, The Open University, Milton Keynes, England

The last twenty years or so has seen an increasing interest world-wide in environment problems, and during this period many countries have introduced legislation or regulations governing the use of certain materials or the implementation of specific trade practices. Nowhere has this intervention been more apparent than in the packaging sector.

However, although the intention of such intervention is to limit deleterious environmental effects, when the actions of different countries are compared, it is often difficult to discern any common objective. In part, this arises from the different trade practices in different countries but it is also due to the very different perceptions of what constitutes an environmental problem. Some countries, for example, see litter as a major problem; others regard it as a minor irritant. In addition, there is the added complication of what is politically acceptable – a factor which frequently leads to totally different solutions to apparently identical problems.

Nowhere is this diversity of approaches to environmental problems more marked than in Europe where, in addition to the problem mentioned above, there are the very different temperaments of the different nations and the limitations on international communications imposed by the language barrier. Despite the existence of the EEC which might have been expected to propose a unified approach, individual countries have pursued their own routes. Not surprisingly, it is difficult to follow developments and the reasons behind them.

A little while ago, a group of European scientists, industrialists and government workers came together at Sjöbo in Sweden under the auspices of the TEM Group of the University of Lund to discuss public intervention in the packaging sector of industry. The aim was to find out what was going on in the different European countries, to see why it was going on and to consider the success or otherwise of the schemes that had been introduced. Each of the participants presented a formal paper which was discussed by the group as a whole. The papers collected together in this book represent the information presented to the conference, amended where necessary to take account of the group discussion. It is important to recognise that the contributors were acting as individuals and not as representatives of their parent organisations.

The papers fall into three main groups. First there are two papers of a more general nature which consider the need for environmental legislation (Boustead) and the basis on which any such legislation can be appraised (Fisher). The second group of papers can be regarded as survey papers in that they attempt to summarise practices in different countries across all types of packaging to which intervention has been applied. One paper (Lieben) summarises actions in the various European countries whilst others look more closely at practices in Denmark (Garmin), Norway (Holdt), Finland (Hämäläinen) and the Netherlands (van Ewijk).

The third group of papers was concerned with selected specific aspects of intervention. Thus, one paper (Cohadon) reported the experiences of the government–industry contract in France. A second pair of papers considered the lead-up to the demand for intervention in the aluminium can market in Sweden (Backman) and the way in which the problem was eventually dealt with (Lidgren). The final set of papers was concerned with milk packaging in Germany (Wietbrauk) and in Holland (Duyvestijn, van Doorn and Middelveldt).

Although in one sense these papers only give a 'snapshot' of the views of the participants in the middle of 1982, it is also true that if this snapshot had been taken at almost any time during the last decade, the result would have been essentially the same, apart from a few minor details. Consequently, although some of the detail may change, the ideas will probably still be current twenty years or more from now.

CHAPTER 2

The physical basis of restrictive legislation

I. BOUSTEAD
Faculty of Technology, The Open University, Milton Keynes, England

---

INTRODUCTION

Over the last decade or so, a number of countries have followed Oregon [1] by introducing legislation regulating the use of certain types of packaging systems and some countries are still considering whether to enact similar legislation. Much of this legislation has been prompted by environmental or conservation considerations and has been aimed principally at limiting the use of non-returnable container systems for beverage packaging. However, if the arguments which inspire such actions are pursued, it is inevitable that similar rules must eventually be applied to other packaging systems.

The use of law in this field seems to indicate that a specific problem and its causes have been identified, that a unique solution has been found and that law has been invoked because it is highly unlikely that the proposed solution would be implemented without some form of legal sanction. All of these assertions are questionable, and it is possible that the long term impact of such regulations may be environmentally undersirable. It is also doubtful whether the short term benefits will be as expected. It is therefore important to consider whether restrictive legislation is appropriate or desirable.

These problems and their implications should be discussed well before any legislation is introduced because an unfortunate consequence of implementing or proposing the use of law in the packaging area has been a polarisation of views. Proponents have come to be regarded as trendy environmentalists and opponents as industrialists with vested financial interests. Both sides are therefore defending entrenched positions and arguments put forward by either side are regarded as suspect.

One disturbing feature of this polarised debate is the ability of both sides to marshal an array of arguments based on a variety of factors such as economics, environment, conservation, pollution, labour, etc., which entirely substantiates their viewpoint. This must be unique in industrial practice, since seldom, if ever, do all factors dictate a single course of action. In addition, many of the contradictory arguments from both sides are based on the same starting data. Clearly both sides cannot be entirely right; nor can they be entirely wrong.

The truth must lie somewhere between the two extremes but whatever action is taken, it cannot expect to satisfy all criteria. It is therefore vital to identify any potential detrimental side effects because they must not create worse environmental effects than those which the law is supposed to be rectifying. Furthermore, if detrimental side effects are too serious, it is questionable whether the course of action should be taken at all. It is the unwillingness of proponents and opponents to spell out the undesirable side effects of their positions that is most disturbing.

This paper is concerned with these doubts and raises a number of questions which should be considered by both proponents and opponents of legislation. Furthermore, politicians and their advisers would do well to probe more deeply into the arguments than they seem to have done until now; in particular, they should question whether they have properly identified the goal they are ostensibly seeking.

ENVIRONMENT AND CONSERVATION

The arguments which demand conservation of raw materials and fuels have been well rehearsed elsewhere [2–5] and need not be detailed further here. The general conclusion however is that the wasteful use of resources should be eliminated and, wherever possible, existing use patterns should be geared to an overall reduction in resource consumption consistent with efficient industrial practice.

This general aim does not conflict with the underlying philosophy of industrial organisations. Most companies welcome a reduction in their overall resource consumption as a means of reducing production costs. The brake however is usually economic, and new production methods tend to appear only slowly as replacement becomes necessary. One problem that conservationists meet is their desire for rapid change irrespective of the immediate effects on industry.

It should also be noted that conservation is concerned with physical quantities of resources; it does not refer to market value. This is not to suggest that economic arguments are unimportant, merely that they are different. A confusing feature of many conservation arguments is the way in which economic and physical data are mixed. When discussing the availability of physical resources, time-independent physical units (e.g. kg, MJ, litre) must be used.

An additional advantage of using physical units is that they are directly related to the technology employed within the system, and whilst this will change with time, the rate of change is very slow compared with the time value of money or market value of commodities which do not depend upon technology. For example, the market value of gold fell from $400 to $313 per troy ounce between January and March 1982; neither the technology for producing gold nor the known gold reserves changed during this period. Similar price fluctuations occurred in other materials, notably oil.

Resource conservation is a widely accepted general aim but it cannot be practised until a set of specific objectives have been defined. As a first stage towards specifying objectives, consider the nature of the systems that are to be modified.

RESOURCE SYSTEMS

A system is defined as a set of operations which perform some specified function. In general, it is not necessary to know how the individual components of the system or the linkages between them work in order to describe the behaviour of the system as a whole. However, it is important to know how the system will respond to external influence, and for physical, man-made systems this frequently implies knowing something of the way the component sub-systems and the links between them operate.

Until the mid-1960s, production technology was primarily concerned with relatively small systems for which the function could usually be specified as the production of some commodity or saleable component. Indeed the engineer was usually concerned with even smaller systems; machines or factories which are only small sub-systems within the much larger production system. Because of this concentration of effort on component sub-systems, the detailed performance of these constituent parts is well understood, and within company organisations, the performance of groups of machines to form factories is often well known in physical terms – even though this information may not be widely known outside of the company for obvious commercial reasons. In all of these type of systems, the emphasis is on describing systems in terms of the output of some physical product.

This emphasis on systems specified by physical output, is carried over into economics. Companies are primarily interested in monitoring their performance in economic terms since this is how their efficiency is judged. Consequently performance data are usually collected in this form. Similarly, large economic systems are usually analysed into sub-systems which correspond to industrial sectors identified by product. Thus company data and national or trade statistics are compatible, and their collection is relatively straightforward. Despite the amount of data available, the behaviour of these product-oriented systems cannot be predicted with absolute certainty but since they have been studied for decades, their general characteristics and, in most cases, the influence of external constraints, can often be inferred with a fair degree of confidence.

In contrast, the systems that must be used in resource analysis differ in three important respects:

(a) they are specified in terms of function *not* product,
(b) they cut across the boundaries of a number of often unrelated product-oriented systems, and
(c) they involve the consumer.

Consider each of these differences in turn because they highlight some of the fundamental problems inherent in proposing conservation measures.

(a) System specification by function

Any materials-processing operation can be represented as shown in Fig. 2.1(a) where the box represents the system boundary enclosing all operations of interest, and the arrows represent inputs and outputs of materials and fuels. Not all of the outputs are necessarily useful or saleable but the system is usually designed to yield a main product or group of products and its performance is usually monitored in terms of the yield of these products.

In the late 1960s, interest arose in the production of products from raw materials in the ground, and the system must be modified as in Fig. 2.1(b) so that inputs are traced back to the earth. In the systems of Figs. 2.1(a) and (b) specification of materials flows may be inadequate to identify the system completely when alternative processing routes are available. For example, steel could be made from iron ore via the blast furnace/basic oxygen furnace or via direct reduction/electric arc furnace. When such alternative routes exist, the precise route must be specified.

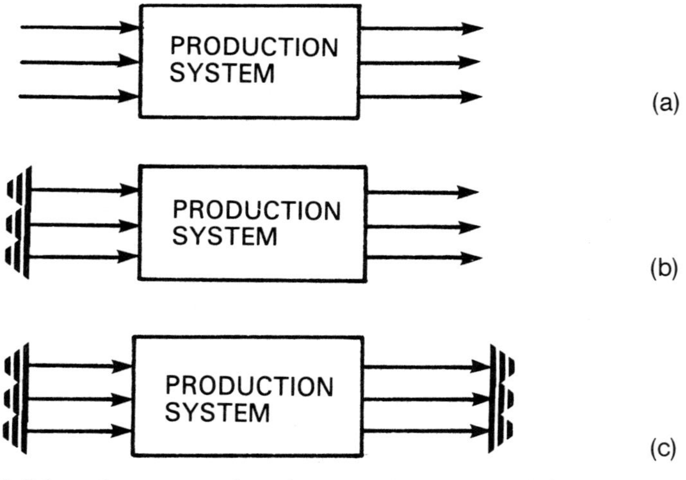

Fig. 2.1 – (a) Schematic representation of a processing system. (b) Schematic representation of a processing system when all inputs are raw materials in the ground. (c) Schematic representation of a production and use system where raw materials inputs are derived from the earth and the only outputs are waste materials returned to the earth.

When systems are defined to describe the production *and use* of a product, system definition in terms of output product is no longer possible. Inputs to these systems are derived from raw materials in the earth, and the only outputs are the ultimate disposal products returned to the earth as shown in Fig. 2.1(c). Flows of waste products are seldom measured, and system function must be defined by some internal function of the system, such as the delivery of a specified amount of product to the consumer. Such systems can be referred to as function-oriented systems to distinguish them from the product-oriented systems described earlier.

(b) Function-oriented and product-oriented systems

Perhaps the greatest conceptual difficulty in resource analysis is the distinction between product-oriented and function-oriented systems. One representation that may be helpful is shown in Fig. 2.2.

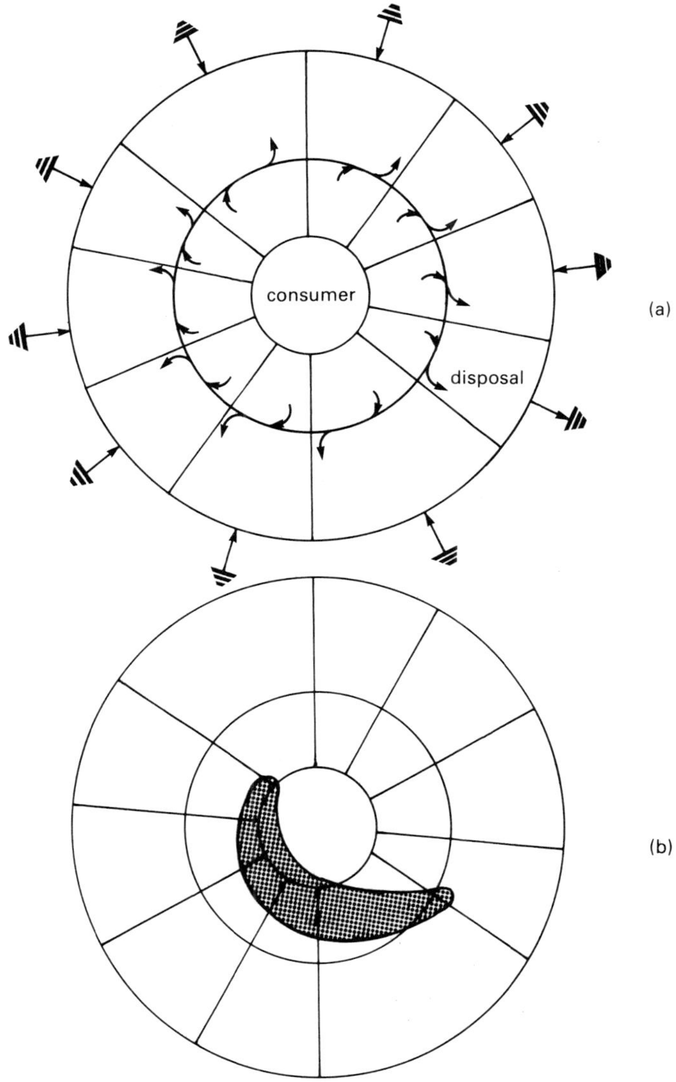

Fig. 2.2 – (a) Schematic representation of industrial activity analysed into product-oriented sub-systems. (b) Function-oriented system (shaded) superimposed on the product-oriented diagram.

Consider first the product-oriented system of Fig. 2.2(a). If the total resource requirements of industrial activity are represented by the outer circle and the central core represents the consumer, then the remaining area can be divided into sectors, each of which corresponds to some product-oriented industrial system. Figure 2.2(a) shows only a few sectors but if, for example, the UK Government trade statistics were used, there would be 155 product-oriented sectors [6].

Each sector draws resources from the earth, process them and either feeds them directly to the consumer or sends them via the central path to one of the other industrial sectors. The consumer after using the products, feeds them into the disposal sector where they either return to the earth (disposal) or are recirculated to one of the industrial sectors (recycling).

Now consider a function-oriented system. Suppose that the function-oriented system to be considered was the delivery of butter in aluminium foil packs. The principal product-oriented sectors involved are the production of aluminium foil, the production of paper and board, transport, retailing, production of petrochemicals, the consumer and the fuel-producing industries. Only a small portion of each sector would be involved, so the function-oriented system, when superimposed on the product-oriented systems, might be as shown by the shaded area of Fig. 2.2(b).

Whereas the industrial activity of most countries has been analysed into product-oriented sectors, so that the whole system of Fig. 2.2(a) can be synthesised from these separate sectors, the same is not true of function-oriented systems. Apart from the beverage container system which has received an inordinate amount of attention in most countries, and one or two other small systems, the analysis of most national industrial activities into a complete set of function-oriented sub-systems has not even been attempted.

The reasons for this are not difficult to discover. First, the choice of a set of non-overlapping systems is extremely difficult, although with the present state of knowledge (or ignorance), a significant number of systems could be defined before this constraint becomes noticeable. It would seem sensible to draw up a list of consumer goods, and examine the systems responsible for their operation, since, as will be seen later, any national conservation policy must identify the heaviest consumers of resources.

The second problem relating to function-oriented systems is that even when they have been defined, the availability of data is very limited. Most readily available data is based on product-oriented systems, and although function-oriented systems can be analysed into a sequence of product-oriented elements, these elements are usually much more specific than those for which data are usually collected. For example, one sub-system in the aluminium can beverage system is the production of aluminium sheet. The most readily available data relate to the production of aluminium rolled products of which sheet is only one; foil, which is much more resource-intensive, is a more significant contributor to this industrial sector. Thus, even when a satisfactory set of systems has been defined, an enormous amount of work remains in obtaining the essential minimum data from primary sources to describe the systems quantitatively. However, if sensible and comprehensive conservation policies are ever to be worked out, these problems must be overcome and the detailed calculations must be performed.

(c) The consumer

The third novel feature of function-oriented systems is the need to include the ultimate consumer as an industrial operation. Figure 2.2 shows that the consumer is central to all processing systems and is also, incidentally, one of the highest direct energy consumers. He also acts as a direct link between the production systems and the disposal or recycling systems. His actions therefore dictate the effectiveness of many systems and are often crucially important as, for example, in returnable container systems.

The problems of measuring the resource requirements of a consumer, and of partitioning these requirements amongst the many functions performed by him, represent an almost impossible task. It is hardly surprising therefore, that most resource analyses tend to omit the resource requirements of the consumer. However, it must be remembered that this is an omission for the sake of convenience and does not correspond to physical reality. Where such calculations have been attempted, the effect is shown to be significant. For example, in the case of the beverage container systems, the consumer acts as the final delivery stage and because the 'load' is small, the specific energy consumption per unit product is high compared with bulk distribution systems. Omission of this contribution from container systems represents a significant error (up to 20%) [7] in the description.

Without underestimating the problem, the resource consumption of the consumer can be treated initially as comprising two parts. The first part is due to all those actions he is required, by the very design of some external systems, to perform. For example, the beverage container systems require the consumer to act as the final link in the distribution chain, and some recycling schemes require the consumer to return materials or goods to a collection point.

The second part of the consumer's resource consumption arises from all other operations: those which he is obliged to participate in, but which cannot be identified, and those which he chooses to participate in. In the initial stages of a comprehensive analysis, it is perhaps wiser to set up a system which covers *all* of the actions of the consumer and to evaluate his total resource requirements. Subsequently, the total requirement can be reduced whenever participation in some other scheme is identified, and a proportion of the total can be ascribed to this other system.

It is important to realise that the resource demand of the consumer is high compared with many function-oriented industrial systems or group of systems. For example, the total consumption of primary energy by all beverage container systems in the UK in 1977 was of the order of 1% of total national consumption of primary energy [7]. In comparison, the primary energy due to consumption of electricity alone by the consumer for purely domestic purposes was some 12% [8] of the total national primary energy consumption. Clearly, serious resource conservation measures must look very closely at the consumer to see where savings can be made. It seems very strange to introduce regulations aimed at reducing the 1% due to all beverage container systems whilst leaving the consumer operation untouched.

DEVELOPING A CONSERVATION POLICY

If any Government is serious about conservation it must decide upon an overall conservation policy at the outset; the introduction of isolated pieces of legislation aimed at some very vague and imprecise goal, achieves little and may well be counterproductive. An overall plan which determines priorities and into which individual schemes can be dovetailed, allows sensible long-term schemes to be devised.

At the present time no country has a comprehensive conservation policy justified on a resource basis even though most governments claim to be in favour of conservation measures.

It would seem therefore that a number of actions should be taken by all governments before they seek to interfere in any way with existing industrial systems which, with all their faults, accomplish the task for which they were set up. In outline these actions are:

(a) to obtain the basic resource data needed to describe all of the function-oriented systems operating at present;
(b) to decide which factors should be used as a measure of resource implications because it is a reduction in these factors which dictate the success or otherwise of any conservation measures;
(c) to determine the priorities and levels of resource reduction that are to be aimed at – using the parameters chosen in (b) above;
(d) to consider the various ways in which the systems could be altered in order to achieve these savings and to choose the most appropriate method;
(e) to consider whether the changes should be induced using regulatory legislation or whether they would be more effectively accomplished by other means; for example by making grants available to stimulate industry to make the changes;
(f) to monitor any schemes set up to determine whether the expected changes are in fact occurring;

Consider each of these proposals in turn because they indicate the essential arguments that must be faced in determining any long-term strategy.

(a) Acquiring basic data

Any changes that are thought to be desirable have to be made in existing systems.

It is therefore essential that an accurate, quantitative description be available for the performance of these systems as they are operating at present since this represents the baseline from which any improvement must be measured. Accomplishing this is a two-stage process. First the function-oriented systems must be identified and defined in some detail and secondly, a quantitative description must be made of each of the systems.

Systems identification may initially appear straightforward but the number of such systems to be defined is very large. For example, a recent analysis of the energy requirements of the UK beverage container systems [7], identified a total of 46 distinctly different systems, and when minor variations in these systems were taken into account, the number increased to over 200. When it is remembered that these systems account for only 1% of total UK energy consumption it is clear that a comprehensive set of systems presents a formidable task.

In the calculations, it is tempting to aggregate data to simplify the task but average values are of only limited use, and certainly cannot be used to determine specific actions in specific systems. It is important in this type of work to discover the variations in practices from one factory to another. For example, in 1978, UK glass container factories were operating with energy requirements varying between ± 19% about the mean value [7]. Changes aimed at improving the performance of these factories, so that they all approached the greatest energy efficiency, would be a more effective way of reducing energy consumption in the glass container industry than tampering with the remainder of the system. Such a solution could not be identified without a detailed analysis and is impossible if data are expressed as average values.

There is also a temptation to import data from other countries. For example, there have been numerous attempts to apply the results of American analyses of beverage container systems to Europe. It is not difficult to demonstrate the futility of this. Distribution and retail patterns in the USA are quite different to those in, for example the UK and even within Europe, there is no standard pattern for these operations.

Great care must also be taken to ensure that any conversion factors are correct A significant example arises in energy analysis where the primary energy required to produce electricity varies widely from one country to another. In the UK for example the efficiency of electricity production and delivery from primary fuel is of the order of 25%; in West Germany a more appropriate value is 31%, and in Norway where a high proportion is hydro-electricity, a value of 80–90% is more appropriate. Obviously, the conclusions to be drawn for conservation of electricity resources in these different countries will be different in the three countries. Thus, a rigid common European policy is unlikely to be a sensible goal when considering specific schemes; instead, the policy must be sufficiently flexible to accommodate regional variations.

There is also a tendency to simplify the interpretation of results by aggregating data. This is particularly true for energy data where all fuels can be converted into a common unit and added together. However, as will be seen later, it is important to ensure that the consumption of different resources be kept separate because of the need to reduce consumption in selected areas. This same approach is needed in monitoring the effects of changes because a reduction in one parameter can often lead to an increase in another.

(b) Choosing parameters

One decision that must be taken at the outset is to decide which resource parameters any proposed measures should seek to influence. There is no single parameter than can be used as an overall measure of the resource implications of any defined system. Basically, there are two groups of resources that are of special interest; fuels and raw materials. Fuels can be treated separately even though measured in a common unit (e.g. MJ) or they can be aggregated to give a gross energy requirement. Occasionally, it should be added, they are treated as raw materials and measured in terms of mass. In contrast, raw materials cannot be aggregated and must be expressed as a list of individually identified inputs.

However, all fuels and raw materials are not of equal importance. For example, it may be argued that, because oil reserves are smaller than known coal reserves, conservation of oil should be given a higher priority than conservation of coal. The problem with raw materials is even more acute. How for example should bauxite and cassiterite be compared?

Thus, an essential precursor of conservation policy is ranking fuels and raw materials in an order of priority for conservation, because changes in systems do not result in a uniform reduction of all the resources consumed by the system. To date no such ranking has been attempted.

It is important to recognise that many arguments proposed for system changes use gross energy requirement as the comparison parameter. To a great extent this has arisen because gross energy requirement is one of the few parameters that has been calculated for many production systems, although it is also a measure of heat release into the atmosphere. However, the fact that it is one of the more easily calculable parameters and that some data are available, should not be taken to imply that it is, in some way, a parameter of supreme importance. Any decision to use gross energy requirement as a resource conservation parameter in preference to other parameters must be very soundly justified. So far this has not happened.

(c) Choosing priorities and levels of reduction

One consequence of evaluating the resource requirements of the set of function-oriented sub-systems is that the sub-systems can be ranked using the criteria chosen in (b) above in the order in which they consume resources, and hence the order in which they should be examined for change. Furthermore, the degree of change to be induced must relate to the savings that are to be achieved; this is a subjective judgement but decisions must be made before changes are implemented because there are few instances where complete elimination of consumption is possible. Where several parameters are to be reduced, the question of priorities becomes even more important.

Deciding the levels of reduction is important not only to the operators of the systems who must be aware of the level of savings they are expected to achieve, but also as a prerequisite to the essential modelling exercise which must follow any proposal for change. Most industrial systems are interlinked, and disturbance to one part of the network inevitably means some readjustments in the rest of the system. It may be that some of these readjustments are unacceptable. For example, suppose that the use of tinplate in packaging were eliminated or drastically reduced. The rolling plant producing sheet for tinplate may also be producing sheet for the motor-car industry. If the demand for tinplate fell, then the production of car body sheet may no longer be possible with the plant running at reduced levels of production. There would, therefore, be a direct influence on the completely unrelated car industry.

Clearly then, setting priorities and reduction levels, followed by modelling exercises when any changes have been proposed to ensure that the priorities and levels of savings are reasonably likely to be achieved, is a vital exercise in resource conservation policy. To date, no such priorities or levels of reduction have been spelled out in any country, and there have been few satisfactory modelling exercises.

(d) Modifying systems

Three types of change can lead to reductions in the resource requirements of industrial systems. First, and most obvious, is the general reduction in resource demand by product-oriented systems. This has been the goal pursued by most industries for decades. In part, such reductions are achieved by developing new or improved processing systems, and this is essentially the job of the engineer. An additional contribution can be made by improving the general 'good housekeeping' aspect of industrial practice. The former savings will probably continue but the latter, once achieved, cannot be repeated, although it must be added that all of the savings in this area that could be made have not yet been achieved.

The second approach is to modify the system internally. Within the beverage container systems, the lightweighting of glass containers is a good example of improvements that can be achieved by the action of a single operator within the system. The area where few advances have been made, however, is the development of schemes involving a number of operations within the system. For example, glass bottles leaving a production line are sterile. If they can be packed to retain this sterility during delivery to the filling plant then there would be no need to hot-wash bottles at the filling plant. It follows that the provision of the most suitable packaging materials would be the responsibility of the glass container manufacturer whereas the saving would occur in the filling plant. The system as a whole would, therefore, be more energy efficient, but the benefits would not be equally spread amongst the participants. One area in which government could act would be to promote such schemes in a way that ensures the benefits are distributed fairly.

The advantage of promoting schemes which *modify* systems internally is that they can be tested in a small way initially without drastically affecting the operation of a system which is known to work. If the scheme does not achieve the expected savings, or if it proves unworkable for any reason, then it can be dropped and the system will revert to its original method of operation.

An extreme example of internal system change arises when a system occurs which cannot apparently be modified in any way and its resource requirements are demonstrably excessive compared to other systems which perform the same or a similar function. In these circumstances, it may be decided to limit the use of the system or, in the extreme case, to prohibit it, relying upon the comparable, competing systems to take over the functions it was formerly performing. Such an action may seem straightforward but there are two problems.

First, as a result of limiting the use of a system, the system may not be able to operate at a reduced level and so would collapse. In the case of prohibition it would disappear anyway. If subsequently it proves that this action was wrong then it is not possible to return to the original state.

The second problem is that there is no guarantee that the competing system, which is expected to take over the function, will continue to operate as it was before the change. The arguments leading to this type of change are seductive but the results are not yet proven.

Suppose, for example that a returnable and an equivalent non-returnable container system operate side by side. Suppose further that it can be shown on specified resource grounds that the non-returnable system was the more resource intensive. Suppose further that the non-returnable system is prohibited, the assumption being that the returnable system will take over the function of the non-returnable system. However, the efficiency of returnable container systems is normally very sensitive to the trippage rate. If, as a consequence of the action, the trippage rate for returnable containers falls, then the eventual effect could well be no change in resource demand or even an increased resource demand. Such limiting or prohibitive regulation should therefore only be used very sparingly and even then, only when the expected result can be shown to be inevitable.

It must be recognised from the earlier discussion of function-oriented systems that the consumer is an integral component in packaging systems. There is an understandable reluctance on the part of politicians to try and change the behaviour of consumers, partly because of arguments about 'personal freedom' but also because of the difficulty in monitoring the effect of consumer response. It is, therefore, common for changes to be made in those systems which require specific action on the part of the 'industrial' operators whilst leaving the consumer operations untouched, or tacitly assuming that the consumer will adapt to match any changed practices with which he is faced.

From a resource viewpoint, the consumer is the single most energy-intensive and resource-intensive part of most function-oriented systems, and if significant reductions in resource requirements are to be achieved, then politicians must overcome their reluctance to interfere and must act directly against some consumer practices. The unlimited use of private transport is one such area. In many countries, the decline of efficient, cheap public transport enforces the use of private transport. If governments are to seriously tackle resource problems, then they cannot ignore the result of their own actions, and many apparently unrelated systems are directly relevant to packaging systems. This is a consequence of the interlinked nature of function-oriented systems.

A third way in which systems can be modified is to leave the existing system essentially untouched but to add on a recycling loop. It is axiomatic that any materials recycling loop will reduce the raw materials demand of the system. The use of recycling loops to achieve energy savings is more complicated but in general, the energy used to produce recycled materials is usually less than that needed to produce the equivalent primary materials. Moreover, with reclamation of fuel bearing materials (such as paper, plastics and wood products) the energy efficiency can usually be greatly enhanced.

It should be noted that the relative efficiency of most returnable container systems is due to the existence of a recycling loop within the system. It is, therefore, surprising that advocates of returnable container systems do not carry their enthusiasm over into support for materials and energy recycling schemes. It has to be added that reclamation and recycling schemes, particularly when applied to recovery from post-consumer waste, recover resources from all function-oriented systems which produce waste.

Given the number of possible options for action, it is somewhat sad, in view of the relatively small amount of quantitative information available, to note that the most of the legislative action taken to date has opted for the somewhat extreme course of limiting or prohibiting the use of certain container systems.

(e) The role of legislation

Laws regulate the actions of individuals or corporate bodies in order to protect, preserve, or promote the well-being of the majority. Good laws are those in which the ultimate goals and the method of achieving these goals meet with such widespread approval that almost all members of society willingly accept and obey the letter *and* spirit of the law. Bad laws are those in which a substantial minority dispute either the ultimate goal and/or the way of achieving the goal. Since conservation law requires the willing compliance of almost the whole of society, it is vital not only that it is a good law, but that it is clearly demonstrated to be good law.

It is tempting to discuss in detail the role that law should play (if any) in conservation and environmental matters because the need for legal sanctions requires as much justification as do the ultimate environmental and conservation goals and the methods of achieving these goals. However, comment is restricted here to listing a number of points that should be considered in the light of the earlier discussion when proposing the use of the law.

(i) The ultimate harm that is to be prevented, or the benefit to be promoted, must be clearly demonstrated.
(ii) Since law relates to the regulation or promotion of specific acts, the direct link between the acts with which the law seeks to interfere and the accomplishment of the ultimate goal must be demonstrated.
(iii) If a number of different courses of action are available for achieving the ultimate goal, but the law is implementing only one, then the reason for choosing this single route must be explained fully.
(iv) It has to be shown that the law is being involved bacause no simpler or more effective process will lead to the desired consequences. This is important because of the law is expensive to adminster and is frequently difficult to enforce.
(v) It must be demonstrated that the proposed law will achieve the desired overall effect.
(vi) It has to be shown that the proposed law will not introduce any serious injustices. More importantly from a conservation viewpoint, it must be shown that the law will not prevent implementation of other conservation measures which may in the long term be as desirable, if not more desirable, than that being promoted.

Finally, it has to be recognised that law can essentially be of two types, prohibitive or enabling. Prohibitive legislation must be enforced by some government agency. Such laws tend to be expensive to enforce properly and, unless they are universally accepted as good, lead to continuous problems of avoidance and evasion. In contrast, enabling legislation such as the provision of cash grants to promote certain types of act, throws the onus onto the operator to apply for the benefits and may well, in the long term, represent a more effective use of public funds. It is unfortunate that most current legislation is of the prohibitive type.

(f) Monitoring the effect of change

If the aim of all these changes is to achieve a reduction in resource consumption, then it is vital that the systems affected are monitored to ensure that the action taken is achieving the desired goals. It has yet to be demonstrated that large scale intervention will be beneficial, even though some short-term gains can be claimed in the very limited interventions that have taken place so far. It is not, however, clear to what extent other systems have been affected by these changes, since little monitoring exists. This monitoring exercise is equivalent to continuing the initial research so badly needed before any sensible conservation policy can be decided upon, and must be recognised as an essential and integral feature of conservation policy.

Conclusions

At present, the state of knowledge on the resource requirements of function-oriented systems is so sketchy that no comprehensive resource conservation policy can possibly be worked out. If conservation is to be regarded as an integral part of government, then government agencies must initiate serious long-term research in this area to establish a satisfactory description of existing practices and in view of the need to monitor the effects of any changes that are introduced, they must regard this as a long-term and probably permanent commitment.

In view of the very limited nature of the information presently available, and in view of the almost total ignorance of the interactions between the function-oriented systems that make up industrialised society, the use of any form of legal sanction to enforce ill defined goals seems premature and can hardly be justified on any known physical basis. Moreover, claims for savings that result from changes in limited systems must be viewed with caution until it can be shown that the savings are not being made at the expense of other, as yet unmonitored, systems.

REFERENCES

[1] State of Oregon Laws 549.810–459.890 (1972).

[2] National Academy of Science and National Research Council (US), *Resources and Man,* W. H. Freeman & Co., San Francisco (1969).

[3] Meadows, D. H. *et al., The Limits of Growth,* Pan Books, London (1972).

[4] Cole, H. S. *et al., Thinking about the Future,* Chatto and Windus, London (1973).

[5] Ecologist, *A Blueprint for Survival,* Penguin Books Ltd, London (1972).

[6] HMSO, *Business Monitor – Report on the Census of Production,* London; annual publication.

[7] Boustead, I. and Hancock, G. F. *Energy and Packaging,* Ellis Horwood, Chichester/John Wiley, New York (1981).

[8] The Electricity Council *Statement of Accounts and Statistics 1977–8,* London.

The author

Dr Boustead is a Senior Lecturer in the Faculty of Technology at the Open University, Milton Keynes, England. He graduated from the University of Birmingham with a B.Sc. in Physics and subsequently obtained M.Sc. and Ph.D. From 1960 to 1965, he was Physics master at the William Hulme School in Manchester and from 1965 to 1972 was Senior Research Fellow at the Royal Military College of Science in Shrivenham. He joined the Open University in 1972.

CHAPTER 3

A conceptual framework for a social appraisal of the beverage container issue

J. C. D. FISHER
Environment Directorate, OECD, 2 Rue André Pascal, 75775 Paris, Cedex 16, France

INTRODUCTION

Beverage containers are currently an issue of considerable public concern and interest. Legislation relating to this subject has been passed or proposed for consideration by the governments in a number of European, Scandinavian and North American countries.

This chapter first demonstrates the need for a sound framework to analyse this topical and controversial issue, and highlights some errors and shortcomings made by earlier studies. The second section develops a framework for a social appraisal of returnable and non-returnable beverage containers.

THE NEED FOR A SOUND FRAMEWORK

The wide range of alternative policy measures

Lieben's paper (see Chapter 4) surveys current legislation, directives and policy measures concerning beverage containers that have been implemented in various countries. See also reference [1]. This reveals a number of alternative and, in some cases, distinctly different policy measures relating to the beverage container issue. The three principal policy options are:

(a) *No action by governments.*

(b) *A 'voluntary' agreement between government and the beverage industry.*
Can and bottle manufacturers, bottlers and retailers agree that the beverage industry will itself undertake measures to attain certain specified objectives such as:

(i) the level of beverage containers disposed in domestic waste should not increase or should decrease by a certain amount;

(ii) the recycling of glass and metal cans should increase to certain target levels;
(iii) an effective litter control programme should be promoted;
(iv) the amount of beverages sold in non-returnables should be limited;
(v) the design and weight of beverage containers should be improved;
(vi) the returnable system should be improved by, for example:
- promoting standardisation
- increasing trippage
- providing (more) information for the consumers on which bottles are returnable with deposits, where the consumers can buy beverages in returnable bottles, and where they can return the empty bottles
- making it easier for consumers to return bottles.

(c) *Government intervention in the market for packaged beverages.*
Such intervention measures might include:

(i) Bans on the use of certain types of non-returnable containers for carbonated beverages.
(ii) Mandatory imposition of refundable minimum deposits on beverage containers.
(iii) Variations of (ii) such as the legislation in Michigan (USA) which requires a mandatory deposit of 10¢ on all carbonated beverage containers but a lower minimum deposite of 5¢ is placed on 'certified' standard returnable beverage containers which can be re-used by more than one bottler.
(iv) Standardisation of bottle size and shape.
(v) Mandatory requirements that retailers give equal shelf space and choice for the consumer between returnables and non-returnables.
(vi) Mandatory requirements that retailers must accept and refund the deposit on any returned empty container of the same type, brand and size as those sold by him.
(vii) Mandatory identification on a bottle or its label that it is returnable for a specified deposit refund.
(viii) Levying high taxes on non-returnable containers.
(ix) Imposition of a 'litter' tax on major items of litter to finance anti-litter campaigns; such taxes were introduced in the state of Washington (USA) and beverage containers were included within the scope of the tax.
(x) Imposition of product charge on packaging items, including beverage containers, to cover their 'external' waste collection and disposal costs.
(xi) Enacting 'bottle bill' legislation such as that originally passed in Oregon and subsequently in other American states which bans metal beverage cans with detechable ring pulls (measure (i) above) and combines the levying of mandatory deposits on certified and non-certified beverage containers (measure (iii) with mandatory retailer acceptance provisions measure (vi)) and a mandatory identification of the refund value of the bottle (measure (vii)).

(xii) Other combinations of the measures (i)–(xi) above which have also been adopted in various countries, provinces or states.

CONCEPTUAL ERRORS AND SHORTCOMINGS IN EARLIER STUDIES

There are many examples of conceptual errors in the current literature on the beverage container issue, these are summarised in references [2] to [5]. Thus, Pearce and Spence [6] state that in relation to the question of 'over-packing', there is a failure in some papers (e.g. references [7] and [8]) to distinguish adequately between internal and external costs and benefits. In particular, Pearce and Spence state that some authors argue or imply that, on account of the internal benefits and cost savings created by packaging, there does not exist any problem of 'over-packaging'. The Incpen paper [8] argues that consumers already pay for the 'external' costs of the disposal of packaging in household waste on account of the inclusion of waste management costs in the rates paid by householders. However, as demonstrated in reference [1], this payment of waste management costs through the current rating system does not remove the 'external' nature of the costs of disposing of packaging waste generated by individual consumers.

Many studies of the impacts of the beverage container legislation implemented in a specific region, primarily comprise a comparison of post- with pre-legislation situations (e.g. references [9] and [10]). However, as the OECD Report [4] correctly points out, what is really required is a comparision of the policy 'ON' with the policy 'OFF' situations. Furthermore, Stern *et al.* [2] state that some papers (e.g. reference [11]) have drawn spurious conclusions concerning the fall in beer and soft drinks sales that occurred in Vermont following the implementation of beverage container legislation since they fail to give consideration to the other factors such as the weather, a reduction in the number of tourists and so on.

More elaborate impact studies do delineate and compare the policy 'ON' with the policy 'OFF' situation [2], [3], [12]. Nevertheless, many of these elaborate impact studies, along with the earlier mentioned studies of the impacts in those regions where beverage container legislation had been implemented, merely itemise their findings for the various categories of impacts. Thus, in relation to the economic impact categories, Stern *et al.* [2] present data on the impact of beverage container legislation on the profit and employment levels and also earnings levels (i.e. payroll costs) in the beverage industry, along with information on the capital investments required as a result of the legislation and the policies' impact on beverage sales and the costs of filling and distributing packaged beverages. The US Department of Commerce Study [13] details data on the impact of beverage container legislation upon employment levels and the price of beverages and also the capital investments required as a result of the legislation. However, in these and other studies [9], [14], there is a failure to acknowledge and demonstrate clearly the element of overlapping and double counting involved in some of the economic impact categories listed. For example, the impact of beverage container policies on employment levels and the extra capital investment required as a result of a shift towards a greater use of returnables, are some of the internal costs that make up the total internal costs of producing and delivering beverages which, to a certain extent, affect the price paid by consumers. Consequently, a false conclusion could be drawn from this presentation of data on various impact categories where there is a failure to highlight the existence of any element of double counting, and a failure to demonstrate clearly any special characteristics of a particular impact category that merits its separate inclusion in the analysis. Exceptions are references [3] and [4] in which a cost–benefit framework was elaborated and adopted. This subject is examined further later. Therefore, it is essential to develop a sound conceptual framework for a study of the beverage container issue. The remainder of this chapter develops such a conceptual framework and sets out the major factors that it is necessary to consider in the analysis.

A CONCEPTUAL FRAMEWORK

The approach adopted here assumes that government should be considering whether the free market mechanism is currently functioning efficiently where efficiency is defined in terms of the achievement of the greatest social welfare for the whole community – or the attainment of the social optimum.

Therefore, a social appraisal of an environmental issue, such as beverage containers, requires first an assessment of whether there is any divergence between the current position generated by market forces and the social optimum. This determines whether govermnemt should consider intervening. If there is any significant divergence, then the second stage of the appraisal is an evaluation of all alternative policy options. The first part of this section analyses the causes of any divergence between the current position in the beverage container market and the socially efficient optimum position. The second part of this section then draws up some criteria for evaluating the alternative policy measures that have been proposed or adopted concerning the beverage container question.

The distinction between internal and external costs and benefits is of fundamental importance in a social appraisal of any issue. Therefore, it is useful first to provide a definition of the internal and external costs and benefits associated with beverage containers.

Internal costs and benefits are the costs and benefits which the consumer fully perceives and which the beverage producers (manufacturers, distributors and retailers) take into account in making their choice between the alternative containers. *External* costs and benefits arise where the production and distribution to the consumer of beverages in the alternative containers, and the disposal of the containers, create impacts which affect the welfare of other members of society and where there are no traded markets or property rights concerning these impacts. Therefore, the individual producer and consumer do not take these impacts into account in their production and consumption decisions concerning packaged beverages. In the absence of any government intervention, there is no incentive or compulsion for them to do so under the current market system. Consequently, these impacts are imposed on society as a whole and go uncompensated (if they are costs) or unappropriated (if they are benefits).

The current position in the beverage container market

The current position in the beverage container market is the present sub-division of the beverage market between the alternative returnable and non-returnable containers. This occurs at the point where the consumers' Marginal Private Valuation (MPV) of beverages in a particular type of container equals the Marginal Internal or Private Cost (MPC) of supplying beverages in that container.

Assuming that the consumer is indifferent to whether the beverage is in a returnable or a non-returnable container, then the returnable bottle will be chosen (by the beverage producer and consumer) if

$$IC^{B}_{NRB} + IC^{F+D+R}_{NRB} > \frac{IC^{B}_{RB}}{T} + IC^{F+D+R}_{RB} \qquad (3.1)$$

where

RB	= returnable bottle;
NRB	= non-returnable container;
IC^B	= the internal costs related to the actual container (comprising the costs of extracting and processing the raw materials, their manufacture into a new bottle/can and the delivery of the empty new container to the bottler);
IC^{F+D+R}	= the internal costs of filling the container with the beverage and the distribution and retailing costs (NB this item also includes the costs of any supplementary paper or plastic packaging used for the distribution of the beverage containers);
T	= the trippage of the returnable bottle.

Expression 3.1 can be summarised as saying that the returnable container will be chosen if the total internal costs of supplying beverages in non-returnable containers (TIC_{NRB}) exceeds the total internal costs of supplying beverages in returnables (TIC_{RB}). Thus expression 3.1 can be restated

$$TIC_{NRB} > TIC_{RB} \qquad (3.2)$$

where

TIC_{NRB}	= the total internal costs of supplying beverages in non-returnables;
TIC_{RB}	= the total internal costs of supplying beverages in returnable bottles.

Therefore, the current private mix of the packaged beverage market between returnable and non-returnable containers is that which minimises the total internal costs of producing and delivering beverages; that is to minimise ($TIC_{NRB} + TIC_{RB}$).

The assumption that the consumer is indifferent to whether beverages are packed in returnable or non-returnable containers is used merely for illustrative purposes and for ease of exposition. It might be considered somewhat unrealistic since surveys have shown a consumer preference for non-returnables on account of their convenience characteristics. However, the principle shown in expressions 3.1 and 3.2 still holds since the returnable container will now be chosen if $TIC_{NRB} - CCP_{NRB} > TIC_{RB}$ where CCP represents a consumer convenience premium. In other words, the consumer would not choose the non-returnable container unless his valuation of this consumer convenience premium exceeded any extra internal costs of the non-returnable system.

The social optimum in the beverage container market

The social optimum in the beverage container market is the socially efficient sub-division of the market for beverages between returnable and non-returnable containers. This social optimum occurs when the marginal social value of beverages in a particular type of container equals the marginal social costs of supplying beverages in that container.

Making the same assumption as earlier (that consumers are indifferent to beverages in returnable and non-returnable containers), then the socially optimal mix of the packaged beverage market between returnables and non-returnables is that which minimises the total social (i.e. internal plus external) costs of producing and delivering packaged beverages; that is, to minimise

$$(TIC_{NRB} + TEC_{NRB} + TIC_{RB} + TEC_{RB})$$

where TIC_{NRB} and TIC_{RB} are as defined as for equation 3.2 and TEC_{NRB} and TEC_{RB} are the total external costs of non-returnables and returnables respectively.

Causes of divergence between current position and social optimum

It has been demonstrated elsewhere [1], [5], that any divergence between the current market position and the social optimum will be caused by the existence of imperfections in the current market mechanism. Three market imperfections appear to be particularly relevant to the beverage container issue:

(a) imperfections in the current market place, such as imperfect information, which result in a lack of consumer sovereignty,
(b) the existence of externalities,
(c) the consumption of finite resources such as energy, which are considered to be incorrectly priced.

Consider each of these in turn.

(a) Lack of consumer sovereignty

This is concerned with whether consumers' decisions in the market place represent their true preferences. It has been argued that the purchase of a non-returnable may not necessarily reveal a preference for beverages in non-returnable rather than returnable containers *if* the consumer does not have full information and freedom of choice in making their decision concerning which of the alternative containers to purchase their beverages in and, if the container was returnable, in being able to return the bottle. The limitations on consumer sovereignty when buying beverages in the various containers depends upon four main factors:

(i) The availability at retail outlets of beverages in alternative types of containers.

(ii) The comparative advertising levels for beverages in returnables and non-returnables and any effect that this has on consumer sovereignty.

(iii) The information that the consumer receives when he purchases a beverage in a returnable or non-returnable container. This will be affected by the bottlers' and retailers' practices for identifying whether a bottle is returnable or non-returnable and the retailers' policy on price labelling of the beverages (i.e. whether or not they separate the price of the beverage as distinct from the refundable deposit on the bottle).

(iv) Whether there are any significant impediments for consumers wanting to return bottles to retail outlets.

(b) The existence of externalities

The importance of externalities, or external costs, to a social appraisal of beverage containers is effectively highlighted by the minimum cost conditions for the current position and the social optimum in the beverage container market. As discussed earlier these conditions are:

(i) Minimum cost conditions for the current private mix of the packaged beverage market between returnables and non-returnables:

$$\text{To minimise } (TIC_{NRB} + TIC_{RB}) \tag{3.3}$$

(ii) Minimum cost conditions for the *social* optimum:

$$\text{To minimise } (TIC_{NRB} + TEC_{NRB} + TIC_{RB} + TEC_{RB}) \tag{3.4}$$

These two expressions highlight an important point concerning this first stage of a social appraisal of beverage containers. This is that the internal costs and benefits of the alternative containers will be fundamentally important in determining the current position and market shares of the alternative beverage containers. It has been internal factors such as the convenience of non-returnables for the consumer and the extra costs and problems involved with the return process for returnables that have caused the shift that has occurred in the United Kingdom from returnables to non-returnables.

However, it is evident from expression (3.4) that these internal costs and benefits *also* enter into the determination of the social optimum. Thus, the internal costs and benefits cannot be the cause of any divergence between the current position and social optimum for the beverage container market. Therefore, such internal factors are irreleveant for, and should not be included in, this, the first stage of the social appraisal of beverage containers. Nevertheless changes in these internal costs and benefits arising from the various policy measures are relevant and should be included in the next policy evaluation stage.

Comparison of expressions (3.3) and (3.4) shows that the existence of any external costs of beverage containers (terms TEC_{NRB} and TEC_{RB} in expression (3.4)) cause a divergence between the current position and social optimum for beverage containers. The terms TEC_{NRB} and TEC_{RB} relate to the beverage container system's generation of external impacts which are unpriced under the current market mechanism. This includes the following five contributions:

(i) The disposal of the beverage container into domestic solid waste (e.g. via the householders' dustbins) (SW).

(ii) The disposal of the beverage container as litter (L). In this context, litter is defined as material that is not immediately placed in the recognised system of waste disposal and collection (e.g. the householders' dustbins). This material then imposes external costs in the form of the expenditure that local authorities and parks departments have to incur to control litter and, if the litter is not picked up, then in the form of the aesthetic eyesore and public nuisance resulting from the litter that is left on the ground in public areas.

(iii) Air and water pollution emissions and solid waste generated at each of the process stages for the returnable and non-returnable container systems shown in Figs. (3.1) and (3.2) (EP).

(iv) Public health and hygiene impacts arising from the presence of any contaminants in the packaged beverage and from the storage of the beverage containers at the retail outlets (H).

(v) The environmental impacts resulting from the transportation between the various process stages. In particular, the urban traffic congestion caused by the distribution of the beverage containers to the retail outlets (U).

Consumption of incorrectly priced finite resources

It is evident that there are a number of imperfections relating to the market for finite resources which could result in the current market price and depletion rate of finite resources such as energy, and perhaps other raw materials, differing from the socially optimal levels. For a detailed discussion of these imperfections see references [1], [15], [16]. Therefore, this third category of market imperfection should include energy and other raw materials that are consumed at the various process stages of the alternative container systems.

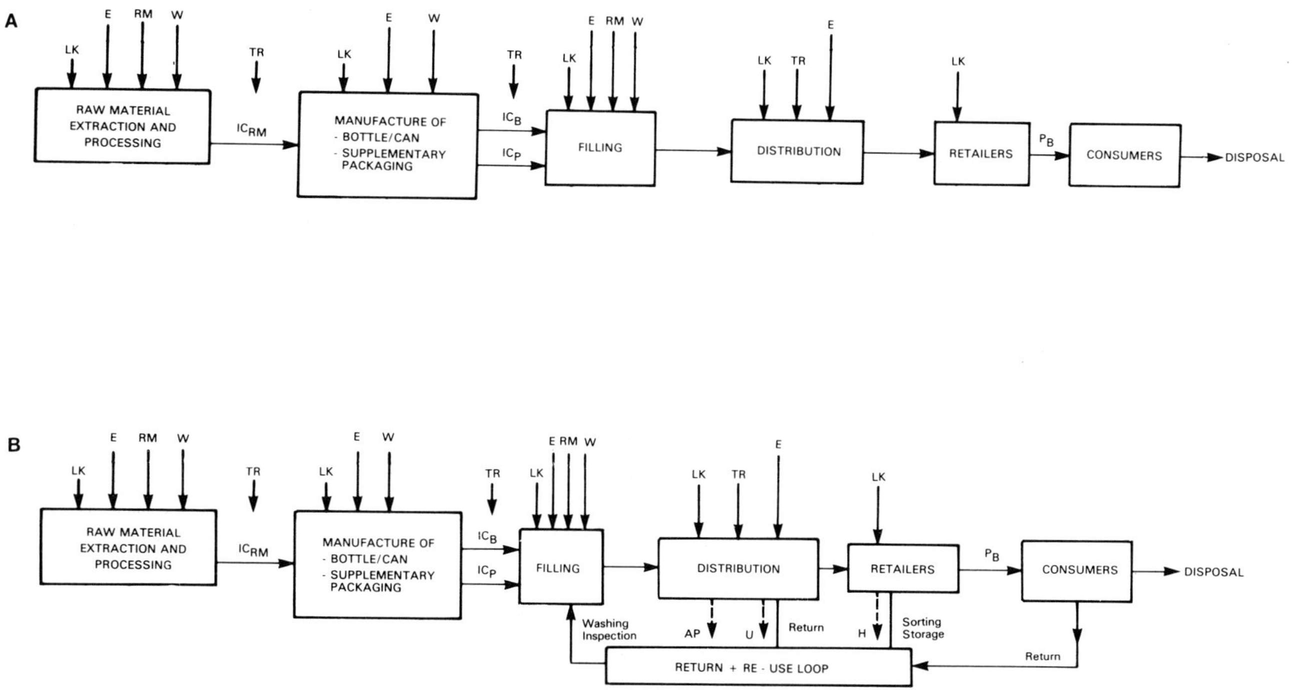

Fig. 3.1 – Process stages in the (A) non-returnable and (B) returnable container systems for carbonated beverages. Internal input categories are LK = labour, capital, land and other factor inputs, E = energy, RM = raw materials, W = water, TR = transport to the next process stage, IC_{RM} = internal cost or price of processed raw materials, IC_B = internal cost or price of the container, IC_P = internal cost or price of supplementary packaging and P_B = price of packaged beverage to the consumer.

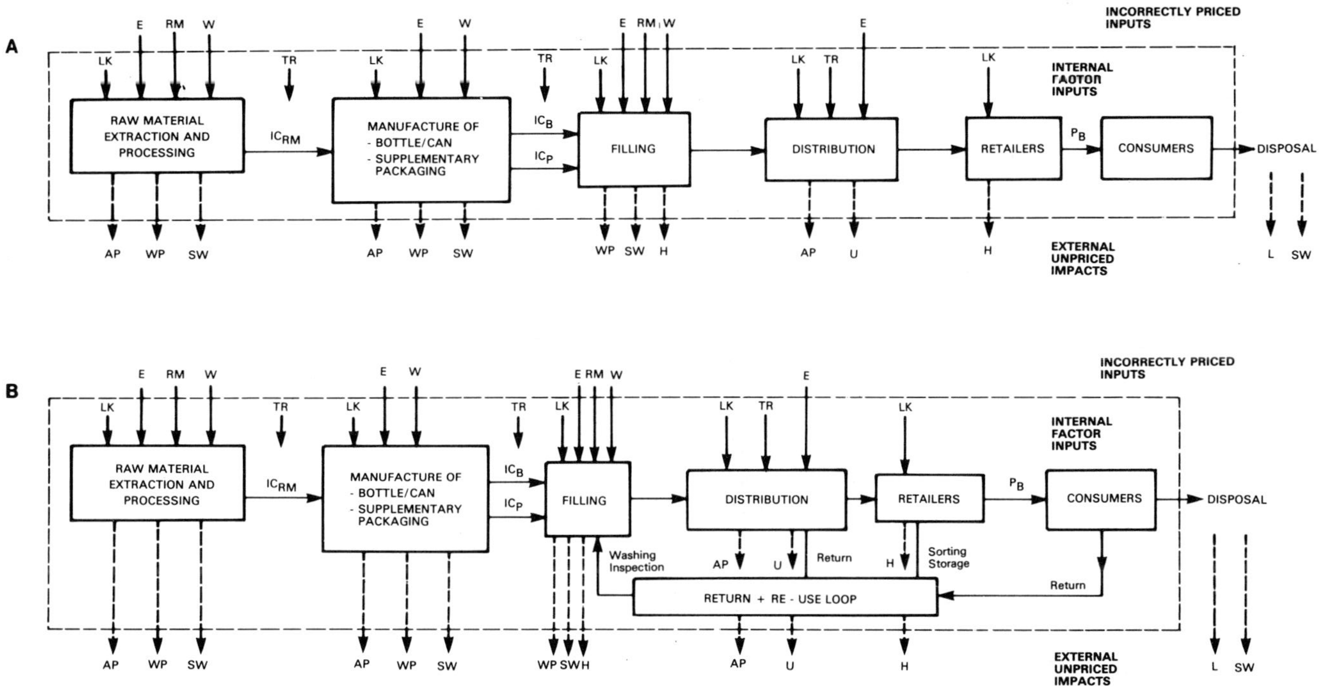

Fig. 3.2 – Process stages in the (A) non-returnable and (B) returnable container systems for carbonated beverages showing internal and external costs. Internal input categories are LK = labour, capital, land and other factor inputs, E = energy, RM = raw materials, W = water, TR = transport, to next process stage, IC_{RM} = internal cost or price of processed raw materials, IC_B = internal cost or price of container, IC_P = internal cost or price of supplementary packaging, and P_B = price of packaged beverage to consumer. External output categories are AP = air pollution emissions, WP = water pollution emissions, SW = solid wastes, L = litter, U = urban traffic congestion and H = hygiene impacts of any contaminants in the packaged beverage and from the storage of the beverage containers at retail outlets.

However, the beverage industry does at present pay for energy and raw materials and these costs are included in the total internal costs of supplying beverages in the alternative containers. Therefore, in a social appraisal of beverage containers, the consumption of energy and raw materials should be weighted by the difference between the market price currently paid by the beverage industry for energy and raw materials (P_E) and the long-run, social opportunity cost of the current consumption of these finite resources (λ_E). The direction and importance of this difference ($\lambda_E - P_E$) will depend upon the net effect of the various imperfections in the market for finite resources.

Figure (3.2) highlights the process stages involved in the distribution of beverages to the consumer in returnable and non-returnable containers. The last two categories of market imperfections are shown schematically in Fig. (3.2) which distinguishes between the internal and incorrectly priced inputs and external impacts for the beverage container systems. The internal factors appear within the box in Fig. (3.2). The resources which might be considered incorrectly priced appear outside and above the box. The external impacts of the beverage container systems are shown outside and below the box.

CRITERIA FOR INTERVENING IN THE BEVERAGE CONTAINER MARKET

The question of whether the government should *consider* intervening in the beverage container market will depend upon the following criteria:

(a) Is there any evidence of the four factors noted earlier resulting in any lack of consumer sovereignty in the beverage market?

(b) Can any significant external costs of beverage containers be shown to exist? This entails determining the size and significance of:

(i) The domestic solid waste generated by returnable and non-returnable beverage containers ($SW_{RB} + SW_{NRB}$).

(ii) The impacts of the returnable and non-returnable beverage containers that are littered ($L_{RB} + L_{NRB}$).

(iii) The pollution emissions and industrial solid waste generated by the returnable and non-returnable beverage container systems ($EP_{RB} + EP_{NRB}$).

(iv) The hygiene impacts of returnable and non-returnable beverage containers ($H_{RB} + H_{NRB}$).

(v) The environmental impacts resulting from the transportation between the process stages of the returnable and non-returnable beverage container systems, in particular, the urban traffic congestion caused by the distribution of returnable and non-returnable beverage containers to the retail outlets ($U_{RB} + U_{NR}$).

These criteria can be summarised in the form of the following expression for the total external costs (TEC) of beverage containers:

$$TEC = \lambda_{SW}(SW_{RB} + SW_{NRB}) + \lambda_L(L_{RB} + L_{NRB}) + \lambda_{EP}(EP_{RB} + EP_{NRB}) + \lambda_H(H_{RB} + H_{NRB} + \lambda_U(U_{RB} + U_{NRB}) \quad (3.5)$$

where

TEC = total external costs of beverage containers
RB = returnable bottle
NRB = non-returnable container
SW = domestic solid waste
L = litter
EP = air and water pollution emissions and industrial solid waste generated
H = hygiene and public health impacts
U = urban traffic congestion
λ = social valuation of these externalities (1)–(5) above.

(c) The requirements of the returnable and non-returnable beverage container systems for finite resources for which the long-run social opportunity cost (λ_E) differs from their current market price (P_E). These criteria can be summarised as:

$$C_E = (\lambda_E - P_E)(E_{RB} + E_{NRB}) \quad (3.6)$$

where

C_E = the extra social valuation (above the market price paid) for the beverage container systems' consumption of finite resources
E_{RB} = consumption of finite resources (e.g. energy and raw materials) in the returnable container system for beverages
E_{NRB} = consumption of finite resources in the non-returnable container systems for beverages
P_E = market price currently paid for the finite resource
λ_E = social opportunity cost of the current consumption of the finite resource.

To determine the direction in which the government should be considering intervention (i.e. favour returnables or non-returnables), it is further necessary to examine three issues. First, whether the lack of consumer sovereignty adversely affects consumers who want to purchase beverages in returnable or non-returnable containers. Second, whether the returnable system generates more or less external costs than the non-returnable container systems. This entails a ranking of the alternative container systems in terms of each of the external cost categories (b) identified above. Third, whether the returnable system requires more or less energy and raw materials than the non-returnable systems.

In relation to the last two issues, for three of the categories of external costs (solid waste, litter, water and air pollution) and for the consumption of energy and raw materials, it is possible and useful to identify separately the external costs and energy and raw materials requirements of the returnable system that arise from the production and disposal of the returnable bottle and hence which are affected by the re-use of this bottle. The relevance of these external costs depends upon whether:

$$\frac{SW_{RB}}{T} \gtrless SW_{NRB} \tag{3.7}$$

$$\frac{L_{RB}}{T} \gtrless L_{NRB} \tag{3.8}$$

$$\frac{EP^{B}}{T} + EP_{RB}^{F+D+R} \gtrless EP_{NRB} \tag{3.9}$$

$$\frac{E_{RB}^{B}}{T} + E_{RB}^{F+D+R} \gtrless E_{NRB} \tag{3.10}$$

where the symbols are as defined earlier.

Expressions (3.7) to (3.10) demonstrate the important effect that the trippage achieved by the returnable bottle can have upon its ranking.

CRITERIA FOR POLICY EVALUATION

In the light of the many alternative policy options listed at the beginning of this chapter, it is essential to establish some definite criteria for the evaluation of the 'best' policy measures. The major points and principles concerning the criteria for the selection of the optimum policy are highlighted here but a more detailed framework of some useful criteria for policy evaluation is given in reference [4].

The objective of this policy evaluation should be to determine which policy measure yields the greatest level of social welfare. To achieve this objective, it is necessary to take into account the environmental impacts and the economic impacts of the various policy options.

The environmental impacts

The enviornmental impacts of a policy include any changes in the level of the external costs of generating of household solid waste, litter, air and water pollution emissions and industrial solid wastes, hygiene and public health impacts, and urban traffic congestion. The environmental impacts of a policy measure also include its impact on the energy and raw material requirements of the beverage container systems.

Economic impacts

The economic impacts of a beverage container policy comprise the impacts of the policy on the beverage industry, the beverage consumer and the government sector. The policy measure may affect the operations, employment and investment levels of the various sectors involved in the process stages of the beverage container systems; i.e. the container manufacturers, the brewers and bottlers, distributors and retailers. Such impacts would result in changes in the total internal costs of supplying the consumer with beverages in the alternative returnable and non-returnable containers. To the extent that this results in changes in the price of beverages for the consumer, the policy's impact on the price of beverages in the alternative containers can therefore be an effective and useful summary measure of the economic impact of the policy. Consequently, changes in the price of packaged beverages have been used as a summary indicator in recent studies on the economic impact of beverage container policies (e.g. reference [3]).

However, a number of factors could result in the measures of the changes in prices failing to capture (fully) certain impacts of the policies. It is also necessary therefore, to take these factors into account in the policy evaluation.

CONSUMER CONVENIENCE AND CHOICE

In addition to changes in the price paid for packaged beverages, the consumer is affected by the convenience with which beverages can be purchased in the various containers and then disposal or return of the container. Frequently, it is currently more convenient for consumers to purchase disposable non-returnables rather than returnables and then returning the bottle. Therefore, it is necessary to take into account the policy's impacts on the *convenience* benefits the consumers obtain from purchasing beverages in non-returnables.

Likewise, it is also necessary to take account of the policy's impact on the *inconvenience* for the consumer of returning bottles to the retail outlets. Surveys of consumer and retailer attitudes and practices concerning the returning of bottles have revealed that consumers in the United Kingdom show it inconvenient to return bottles, primarily on account of the difficulties that they encounter in returning or attempting to return bottles to the retail outlet [17], [18]. Consumers' welfare will also be affected by other aspects of the policy relating to consumer sovereignty, such as the provision of information and the policy's impact on the consumers' freedom of choice. This latter aspect will be influenced by the policy's effect upon the availability at the retail outlet of beverages in the described container size and type (i.e. returnable or non-returnable).

GOVERNMENT PRICE CONTROLS

The existence of any government price controls may mean that the beverage industry is unable to pass any increases in internal costs on to the consumer in the form of increased prices. This means that it will also be necessary to assess the policy's impacts on the profit levels of the beverage industry. Alternatively, it may be simpler to assess the economic impacts of the policy in terms of its impacts on the total internal costs of packaged beverages given earlier as

$$TIC = TIC_{NRB} + \frac{IC^{B}_{RB}}{T} + IC^{F+R+D}_{RB} \tag{3.11}$$

If the policy lowered the total internal costs and hence possibly prices of supplying the consumer with packaged beverages, this would constitute an economic benefit of the policy; conversely, an increase in total internal costs would constitute an economic cost of the policy.

EMPLOYMENT OF LABOUR

The labour costs actually incurred by the beverage industry are included as one of the cost items in the total internal costs function given above in equation (3.11). However in a social appraisal, the employment of labour should be weighted not by the wage rates actually paid, as applied in the internal costs function above, but by its shadow wage (W_λ). A more detailed discussion of this matter is given in reference [19]. This shadow wage is the opportunity cost to society of employing the labour in that job instead of its best alternative job. The high levels of unemployment currently experienced in many parts of the country by certain categories of labour (especially unskilled labour) mean that there may be few alternative jobs available for this labour so that its shadow wage rate is low. On account of social considerations and imperfections in the labour market, however, this shadow wage rate could be less than the wage rate actually paid for these categories of labour. There may also be some instances where, on account of the effect of incomes policies and rigidities in the labour market, the shadow wage for a specific category, (e.g. skilled labour) may exceed its actual wage rate.

Social benefits would therefore accrue from an increase in the employment opportunities for these categories of labour. Recent government measures to increase employment opportunities, such as the activities of the Manpower Services Commission, the Job Creation Programme and the Training Opportunities Placement Scheme, give some support to this view. It is necessary therefore, to ascertain the policy's impacts on the employment levels for the various categories of labour in the beverage industry. These changes in employment levels should then be weighted in the social appraisal by the difference between the shadow wage and the wage rate actually paid by the industry for the various categories of labour so that the appraisal can give an indication of the social costs and benefits of the employment impacts of a policy measure.

CAPITAL OBSOLESCENCE AND INVESTMENT

The policy may result in the obsolescence of certain capital equipment (e.g. filling lines) which will not appear as cost and price changes if the beverage producers adopt the economic principle that 'bygones are bygones'. Thus, although a producer would attempt to raise prices and hence increase his revenue so as to be able to earn an adequate return on any additional new investment, he would not do so for obsolescent equipment which should be written off as an irretrievable capital loss. Therefore, in the social appraisal, any capital equipment rendered obsolete as a result of the policy measure should be included as an additional item, and should be valued at the (social) productivity of this equipment in the absence of the policy measure, less any revenue the beverage producer obtained from the sale of the obsolescent equipment.

OECD [4] also reports that it has been suggested that the impacts upon the beverage industry's investment levels, and hence capital resources, may be of such importance that they merit special assessment. However, it is important to note that a firm's investments are already included in the firm's calculations of its total internal costs (e.g. by depreciation, interest charges, etc.). Therefore, to avoid double counting, the investment impacts should only be included as a supplementary item in the economic impact statement where this importance or social valuation of these investments exceeds the allowance that has already been made in the calculation of the beverage industry's total internal costs. Moreover, in the social evaluation of policies, the impacts of a policy upon a firm's investment should only be weighted by this *excess* of the social valuation of the investment requirements over the allowance for the costs of this investment which has already been made by the beverage industry.

OTHER FACTORS

Other factors that must be included are the impacts of the policy upon the competitive structure of the beverage industry, the distribution of income, the balance of payments, the international trade implications of the policy and the impacts on the government sector. This latter factor includes the administrative feasibility and costs of implementing the policy and the impacts upon government revenues and expenditures. Further information on these other factors is given in the OECD report [4].

SUMMARY

The criteria used to evaluate the various policy options can be summarised as entailing an assessment of the impacts of the policy upon:

(a) The external costs of beverage containers. These impacts are the changes in the function for the total external costs of beverage containers given by

$$\mathrm{TEC} = \lambda_{SW}\frac{(SW_{RB}}{T} + SW_{NRB}) + \lambda_L\frac{(L_{RB}}{T} + L_{NRB})$$

$$+ \lambda_{EP}\frac{(EP^{B}_{RB}}{T} + EP^{F+D+R}_{RB} + EP_{NRB})$$

$$+ \lambda_H(H_{RB} + H_{NRB}) + \lambda_U(U_{RB} + U_{NRB}) \quad (3.12)$$

where the terms are as defined earlier.

(b) The conservation of resources which is represented by changes in the expression

$$C_E = (\lambda_E - P_E)\frac{(E^{B}_{RB}}{T} + E^{F+D+R}_{RB} + E_{NRB}) \quad (3.13)$$

where the terms are as defined earlier.

(c) The total internal costs of supplying the consumer with packaged beverages. These are

$$\mathrm{TIC} = \mathrm{TIC}_{NRB} + \frac{IC^{B}_{RB}}{T} + IC^{F+D+R}_{RB} \quad (3.14)$$

where the terms above are as defined earlier.

(d) Employment levels in the beverage and related industries given by

$$\sum_{i=1}^{n} [\Delta \mathrm{empl}^i (W^i_A - W^i_\lambda)] \quad (3.15)$$

where

i = ith category of labour

n = numbers of categories of labour

empl^i = the policy's impacts on the employment levels for the ith category of labour in the beverage industry

W^i_λ = the shadow wage for the ith category of labour

W^i_A = wage rate paid by the beverage industry for the ith category of labour.

(e) Capital obsolescence and investment requirements of the beverage and related industries given by

$$\mathrm{IV}(\lambda_{IV} - IC_{IV}) + \mathrm{OC}(\lambda_{OC} - NRV_{OC}) \quad (3.16)$$

where

IV = the investment undertaken by the beverage industry as a result of the policy measure

λ_{IV}	= the social valuation or the social opprotunity cost of the use of the firm's capital resources to finance this investment
IC_{IV}	= the firm's internal costs associated with this investment
OC	= capital equipment rendered obsolescent as a result of the policy measure
λ_{OC}	= the social productivity of this capital equipment in the absence of the policy measure
NRV_{OC}	= revenue received by the beverage industry from the sale of this equipment

(f) Consumer convenience, consumer inconvenience, consumer choice and other aspects relating to consumer sovereignty.

(g) Other factors which include: impacts upon the industry's competitive structure; impacts upon the distribution of income, balance of payments and upon the government sector.

Some important considerations concerning these criteria

Two particular factors play an important in determining the economic and environmental impacts of a policy measure. These are the market shares held by the alternative beverage containers and the trippage achieved by the returnable bottle. The other important factors determining the size and nature of the economic and environmental benefits of beverage container policies are analysed in depth in OECD report [4]. This is particularly evident in expressions (3.12)–(3.14) which highlight that these two factors are important determinants of the total external costs and the resource requirements of the alternative beverage container systems. Significantly, they are also two factors which a beverage container policy can alter. In the formulation of a beverage container policy it is therefore essential that the policy should be designed so as to raise trippage.

Finally, for the evaluation of the alternative policy measures it is necessary to have a baseline against which to compare the impacts of each of the policy options. This baseline should be the policy 'off' or 'No Action' case and will be represented by the values for such important factors as the market shares and trippage that would have occurred in the absence of any policy measure.

Acknowledgements

The opinions expressed in this paper are the author's own and so not necessarily represent the views of the OECD.

REFERENCES

[1] Fisher J. C. D. (1982) *A Social Appraisal of the Environmental Impacts of Returnable and Non-returnable Beverage Containers,* Public Sector Economics Research Centre, University of Leicester, Draft Ph.D. Thesis.

[2] Stern C., Verdieck E., Smith S., Hedrick T. (1975) *Impacts of Beverage Container Legislation on Connecticut and a Review of the Experience in Oregon, Vermont and Washington State,* Department of Agricultural Economics, University of Connecticut. Report commissioned by the Connecticut Department of Environmental Protection, Connecticut, USA.

[3] Tawil J. J. (1976) *Social Costs of Beverage Containers: an Economic Analysis of Six Legislative Solutions for Maryland,* Division of Research Department of Economic and Community Development, Maryland, USA.

[4] Organisation for Economic Cooperation and Development (OECD) (1978) *Beverage Containers: Re-use or Recycling,* OECD Paris.

[5] Fisher J. C. D. (1978) A Social Appraisal of Beverage containers Recycling: A UK Perspective *Resources Policy,* Vol. 4, No. 2, June 1978, pp. 78–79.

[6] Pearce D. W., Spence S. (1975) *The Social Costs and Benefits of Packaging with special reference to Returnable and Non-returnable Carbonated Drinks Containers,* Public Sector, Economics Research Centre, University of Leicester. Report on a pilot study for the UK Department of Environment.

[7] Glass Manufacturers Federation (1973) *The Glass Container Industry and the Environmental Debate,* Report by the Glass Industry Liaison Working Party with Government Departments. Glass Manufacturers Federation, London.

[8] Industry Committee for Packaging and the Environment (INCPEN) (1976) *Waste Disposal and Packaging INCPEN Discussion Paper No. 6,* Industry Committee for Packaging and the Environment, London.

[9] Applied Decision Systems (1974) *Study of the Effectiveness and Impact of the Oregon Minimum Deposit Law,* Report prepared by Applied Decision Systems for the Emergency Board, State of Oregon, USA.

[10] Statens Offentliga Utredningar (SOU) (1974) *Effekter Av. Förpackning-savgiften,* SOU 1974:44.

[11] Environmental Information Service (undated) *Effects of Container Legislation on Jobs, Prices and the Environment,* Environmental Information Service Fact File, Seattle, USA; Environmental Information Service (undated) *Questions and Answers about Oregon-type Beverage Container Laws.* Environmental Information Service, Seattle, USA.

[12] Research Triangle Institute (1976) *Energy and Economic Impacts of Mandatory Deposits,* Report prepared for the Federal Energy Administration by Bingham, T. H. *et al.* Research Triangle Park, North Carolina, USA.

[13] United States Department of Commerce (1975) *The Impacts of National Beverage Container Legislation,* Bureau of Domestic Commerce Staff Study, US Department of Commerce, Washington DC, USA.

[14] United States Environmental Protection Agency (US.EPA) (1972) *The Beverage Container Problem. Analysis and Recommendations,* Report prepared by Bingham T. H., Mulligan P. F., Research Triangle Institute

for Office of Research and Monitoring, US Environmental Protection Agency, Washington DC, USA.

[15] Pearce D. W., Rose J. (eds) (1975). *The Economics of Natural Resource Depletion,* Macmillan Press Ltd, London.

[16] Lecomber R. C. (1979) *The Economics of Natural Resources,* Macmillan Press Ltd, London.

[17] Fisher J. C. D., Horton P. (1979) The Return of Returnable Bottles: A Study of Consumers' and Retailers' Attitudes and Practices concerning Returnable Bottles, *Resources Policy* Vol. 5, No. 4, December 1979, pp. 279–297.

[18] Fisher J. C. D. (1981) Returnable and Non-returnable Beverage Containers and Household Waste in Lidgren, K., Bridgewater, A. (eds) (1981) *Household Waste Management in Europe: Economics and Techniques,* Proceedings of an International Scientist Meeting on Household Waste held in Alcudia, Majorca, June 9th–12th 1980.

[19] Mishan E. J. (1975) *Cost–Benefit Analysis* second edition, George Allen and Unwin, London.

The author

Jonathan Fisher graduated from Bristol University with a degree in Economics. He then spent three years undertaking research in environmental economics at the Public Sector Economics Research Centre of the University of Leicester. This research covered the recycling of waste materials, the returnable/non-returnable beverage container controversy, pollution taxes and resource depletion. He assisted in the production of a report 'Beverage containers: re-use or recycling' for the Waste Management Policy Group of the OECD. He has also written a number of other published articles and reports on these environmental economics topics. He is currently working at the Environmental Directorate of OECD where he has been engaged in studies of the source separation of recyclable waste materials, the benefits of environmental policies and the impacts of environmental policies upon technological change.

CHAPTER 4

Measures to promote re-use or recycling of beverage containers

P. LIEBEN
Environment Directorate, OECD, 2 rue André Pascal, 75775 Paris Cedex 16, France

INTRODUCTION

In 1978 the Organisation for Economic Cooperation and Development (OECD) published the findings of a study which the Waste Management Policy Group (one of the working groups of the OECD Environment Committee) has been conducting in the field of beverage containers and solid waste management [1]. The objective of the study was to consider the impact of various beverage container systems upon resource use and the environment, as a basis for government intervention to correct those impacts found to be socially detrimental.

Considering the historical background and likely trends in the beverage container market, the study indicated that unconstrained market forces were generating a steady move towards the use of one-way, non-refillable containers. Experience in most countries suggested that any increase in the market share of the refillable container would only come about through some form of government intervention.

Identification of the impacts of various beverage container systems was achieved through discussion of the relevant external costs imposed by the systems: solid waste generation, litter, air and water pollution, energy use, hygiene and health aspects. It was shown that the refillable beverage container system would, in most cases, generate lower external costs than any other system. Obviously, this finding is dependent upon the trippage achieved by the container (number of times the same container is refilled) and this was considered in great detail. Generally, the necessary trippage was found to be achieved, or even greatly exceeded, in practice.

Possible types of corrective government measures and policy options to cope with adverse environmental impacts were then discussed in terms of their likely success and their possible consequences in respect of other economic considerations such as employment, investment and consumer aspects. The discussion included such measures as mandatory deposits, bans, product charges and various taxes, standardisation, materials recovery and recycling.

Having considered the results of this study, and conscious that measures had to be taken to reverse the recorded trend towards the growth in solid waste generation from beverage container discards, the Waste Management Policy Group drafted the following Recommendation concerning the re-use and re-cycling of beverage containers, which was formally adopted by the OECD Council in February 1978:

The Council:

On the proposal of the Environment Committee;

I. RECOMMENDS that Member countries, through international co-operation as appropriate, where practicable, define and implement policies designed to ensure that the costs of the adverse environmental impacts of the manufacture and use of beverage containers are effectively and equitably borne by the producers and users of such containers.

II. RECOMMENDS that Member countries adopt appropriate measures with a view to maintaining, or where necessary introducing, a system of distribution by refillable containers covering as much as possible of the beverage trade when it is expected that, in doing so, the social costs of the beverage distribution systems are minimised.

III. RECOMMENDS that, when measures to promote the use of refillable beverage containers are considered, they be accompanied by an effort to standardise such containers, possibly undertaken on the basis of collaboration between the countries concerned in order to prevent trade barriers.

IV. RECOMMENDS that, regardless of the measures taken to promote the re-use of beverage containers, Member countries encourage the recycling of the ultimately disposed-of containers, and take any other necessary step to reduce as much as possible any adverse effect they may have on the environment.

Most OECD Member countries have now enacted legislation and adopted practical measures to implement this Recommendation. We shall briefly review them country by country.

MEASURES TAKEN BY GOVERNMENTS

Australia

Waste management and recycling matters in Australia are primarily the responsibility of State and local governments. The Federal Government, together with the states and territories, examine national issues in the waste management and recycling field through participation in the Australian Environment Council (The Council of Environment Ministers).

The State of South Australia promulgated a beverage container act in 1977. This act, which is based on a compulsory deposit and return system, was primarily designed to stem the increase in non-returnable containers and to encourage the use of refillable containers. In practical terms, it prohibits the sale of non-returnable, non-refillable beer and soft drink containers. Containers must all show the refund amount applicable to that container. A study into the economic impact of the beverage container Act was published in 1980 by the South Australian Department for the Environment.

At present there is no other promulgated policy in Australia regarding beverage containers apart from anti-littering laws. It is noteworthy, however, that the federal budget in 1981 exempted all refillable soft drink beverage containers from sales tax.

There is considerable emphasis in all states on recycling of beverage container materials where this is economically viable, particularly glass cutlet, steel and aluminium cans. The return rate for aluminium cans in Australia, in 1981, was around 50%.

As a result of work by the Litter Control Committee (LCC) of the Australian Environment Council, the Council has invited industry to cooperate with Government on various actions for preventing and reducing litter. The report of the LCC is expected to be released in the near future. The Council expressed considerable concern about the continuing trend to non-refillable and non-reusable packaging, particularly in areas traditionally noted for the use of refillable packaging. Ministers endorsed recycling efforts by industry, but at the same time have requested industry to consider implementing the following action:

(a) Review and improve recycling rates for particular materials in the waste and litter streams, such as aluminium, steel, glass and paper.
(b) Consider the payment of maximum resource values for used materials which consumers recycle.
(c) Within marketing strategies, consider the potential for informing consumers of the overall benefits of re-use and recycling, such as lower prices to the consumer.
(d) Identify the cost of packaging to consumers.
(e) Introduce on all consumer goods likely to be littered, conspicuous, innovative and appealing invitations to encourage consumers to recycle or dispose of such products properly after use.
(f) Consider the scope for research and development into more innovation design for disposability and recyclability in new products and packaging.
(g) Consider the possibilities for reversing the trend towards one-way beverage containers, for example through a phased increase in the use of refillable

beverage containers or the introduction of standard refillable bottles which can be used by all fillers and thus maximise the economic viability of this approach to packaging.

(h) Avoid the use of advertising which suggests that particular packages are recyclable where no practical and economic system exists for recycling such packages.

(i) When quoting recycling rates ensure that post-consumer recycling rates are separated from industry 'in-house' figures.

The Australian Environment Council has now established a packaging industry liaison sub-committee to pursue these actions with industry. More recently Council called on industry to replace detachable ring pull tops on beverage cans with an environmentally acceptable alternative on a national basis. Satisfactory alternatives have been used in South Australia for a number of years.

Canada

The beverage container issue in Canada has been mainly considered on a provincial basis since solid waste management is a municipal or regional government function carried out under provincial legislation. Although the federal government monitors developments at the provincial level and has looked at initiatives available at the federal level, there is no direct federal involvement in beverage container regulation at present, except for specification of container size which is the responsibility of the Department of Consumer and Corporate Affairs.

During the 1970s the provinces, one by one, decided to introduce initiatives to reverse the marked trend towards non-refillable, throw-away beverage containers, especially in the soft drinks and beer market. Some of the principal regulations introduced (used to a varying extend and in differing combinations by the provinces) are:

(a) Mandatory deposit-refunds on refillable containers for soft drinks and/or beer.

(b) Mandatory deposit-refunds on both refillable and non-refillable containers, i.e. soft drink and/or beer and/or wine and spirits.

(c) Banning of certain containers, i.e. plastic, aluminium, non-refillable glass, all non-refillable containers and those with detachable openings.

(d) Mandatory availability of refillable containers.

(e) Equal shelf space for beverages in refillable containers.

(f) Taxes on non-returnable containers.

(g) Establishment of depot systems to facilitate the return of beverage containers.

(h) Miscellaneous clauses pertaining to advertising, identification of containers, size of deposits, responsibility of retailers, maximum returns, handling charges, penalties, non-carbonated soft drinks, labelling, vending machines, flavours stocked, standard-sized milk containers.

One of the readily identified overall benefits has been a general reversal of the trend towards increased use of non-returnable containers. It is interesting to note that in most provinces the use of non-refillables continued to increase until government controls were introduced. The one exception to this was in the Province of Manitoba where a co-operative approach with industry was achieved whereby non-refillable glass containers were voluntarily reduced and cans were not actively promoted. As a result, over 70% of the soft drinks sold in this province are in refillable containers. It is also of interest to note that the degree of success, based on certain parameters, has been about the same for several provinces using various combinations of regulatory clauses. For instance, the provinces of British Columbia, Alberta and Saskatchewan have each reached the 80% level of refillable containers although somewhat different initiatives have been used. Ontario is approaching the 70% level with a completely different set and aided to a certain extent by the co-operation of industry after considerable encouragement.

Other general observations on some of these initiatives can be made:

(i) Most provinces have legislated mandatory deposit-refund systems for refillable soft drink and beer containers although the details of the regulations vary considerably from province to province.

(ii) The one province (Alberta) with a mandatory deposit-refund system on wine and liquor bottles (5¢ per bottle) has reduced the quantity of these bottles in the waste stream by 50% compared to the national average. It is thought that a higher deposit (of 10–15¢ say) might further reduce these bottles to a negligible amount. In the provinces of British Columbia and Alberta there is an indication that the mandatory deposit on non-refillable glass and metal soft drink containers has substantially reduced their presence in the waste stream.

(iii) Mandatory availability and equal shelf space for refillable soft drink containers is considered an important requirement by many provinces for increasing the use of refillables. One of the main restraints on the use of refillables has been chain store reluctance to stock beverages in refillable containers as this entails the responsibility for accepting them when returned.

(iv) Bans on certain types of containers and materials used for containers have been used by several provinces with varying degrees of success. For instance, Saskatchewan has banned all non-refillable bottles for soft drink and beer; there appears, however, to be enough exceptions to the rule to permit up to 20% of sales still in non-refillable containers.

(v) The beer industry seems to have maintained a high use of refillables mostly on a voluntary basis encouraged by provincial governments. As mentioned previously, the soft drink industry in Manitoba has also maintained a high use of refillables without government legislation.

One of the more significant recent events in Canada has been the banning on 28 August 1979 of all 1.5 litre carbonated soft drink glass bottles. This action was taken after accidents to individuals from spontaneously exploding bottles continued despite repeating warnings to private industry from the federal government.

As a result of the ban, soft drink bottlers have been attempting to pick up the slack in packaging capacity through the use of the two-litre plastic (PET) non-refillable bottles.

While this container is now being used in three provinces, the other provinces have banned or are resisting its use. The government of Ontario have completed an environmental re-appraisal of soft drink packaging in the province with particular reference to the anticipated impact of PET containers [2]. The request by the soft drink industry to use the PET container in Ontario was rejected on the basis of the findings of this study.

Denmark

Denmark has a well-functioning deposit-refund distribution system for soft drinks and beer based on refillable bottles with a very high return rate. For soft drinks, the use of one-way containers has been prohibited since 1977. For beer, at the end of the 1970s, the use of one-way bottles was practically nil and the use of cans was gradually being phased out through voluntary agreement. In recent years, however, there has been an increase in the number of different types of refillable bottles for soft drinks and beer. This increase was regarded as a threat to the continued efficient operation of the return system. The Agency for Environmental Protection, therefore, proposed a regulation to ban the use of one-way containers and limit the number of different types of refillable bottles for beer and soft drinks. This regulation (Statutory Order no. 397 of 2 July 1981) came into force on 1 January 1982. The ban on one-way containers also covers the PET plastic bottle, insofar as this type of container is not refillable.

For wine and spirits, the use of refillable bottles has seen an upsurge after a long period of decline. Through private initiative, a standardised refillable wine bottle has been introduced which covers a considerable portion of the wine bottled in Denmark. The largest domestic producer of spirits uses a refillable bottle for their aqvavit, which has a major share of the total spirits market. The official policy in this field is clearly in favour of increased use of refillable bottles and government initatives are being considered.

The above-mentioned standardised wine bottle is made in four sizes: 37.5, 75, 150, and 200 cl, which the producers expect will become European or international standard sizes. It is felt, however, that the use of a national standard bottle is only a second-best solution, since a large portion of the wine consumed in Denmark is imported in bottles of numerous shapes and sizes.

Finland

Until very recently, refillable containers have largely dominated the Finnish market for soft drinks, beer, wine and spirits although there was a marked trend towards increased use of non-returnable containers for beer and soft drinks. In 1981 one-way containers represented about 2% for soft drinks (mostly aluminium cans), 8% for beer (entirely bi-metal cans), and 7% for wine and spirits (glass bottles). Milk and fresh fruit juice, however, were entirely packed in non-returnable containers (coated cardboard and plastic).

To stop the trend towards greater use of one-way containers a special tax was imposed on such containers in the beginning of 1976 (0.36 F. Mark per litre for beer, and 0.60 F. Mark per litre for other carbonated beverages). This tax was raised to 1.0 F. Mark per litre at the end of 1980. However, because of the fear that the use of non-returnable containers (especially aluminium cans) might still tend to increase in future, the Parliament decided to raise this tax from 1 to 3 F. Mark per litre on glass and metal non-returnable containers for soft drinks. This new regulation came into force in the beginning of 1982. Similar measures would be taken for other beverages, if the need arises. At the moment, there seems to be an increase in the use of non-returnable plastic bottles for soft drinks.

France

In accordance with the OECD Recommendation on Beverage Containers, the industries concerned signed an agreement with the French Government on 17 December 1979, on measures to protect the environment and to save energy and raw natural resources in connection with the packaging of liquid food.

Under the terms of that agreement, by 1984 the amount of liquid-food containers in household waste is to be reduced by 40%, and the amount of energy consumed per packaged hectolitre is to be reduced by 12% in comparison with 1976, i.e. an energy saving of 150,000 tonnes of oil equivalent.

To achieve these aims, a combination of measures is being applied. The first approach is through technological innovation. Research has been intensified to develop containers which consume less energy for a given quality content.

The second approach has been to develop ways of recovering materials from household wastes. The aim is to recycle 450,000 tonnes per year of household glass and 10,000 tonnes per year of PVC, while 200 million glass bottles are to be re-used.

Lastly, there is to be a gradual return to more economic forms of packaging in cafes, hotels, restaurants and institutions: beers, mineral waters and soft drinks will only be supplied in returnable containers on which a deposit is refunded. Elsewhere, the consumer will be given a choice between returnable and non-returnable containers. In association with the authorities, the industry has undertaken to expand the use of standardised glass containers wherever market conditions permit.

The agreement also provides for the establishment of a monitoring unit by the industry and the authorities. This is intended to provide an annual assessment of progress, and to check on results achieved so that any necessary corrective measures can be taken.

A first assessment of progress made under this agreement was made in 1981, covering results achieved in 1980. It indicates that the energy consumption, per packaged hectolitre, was reduced by 1.2%. This reduction was obtained, in particular, through a decrease of 6% in container unit weight for an equivalent quality and through the separate collection and recycling of cullet. Separate collection schemes were operated by 6,300 municipalities, with a total of 180,000 tonnes of cullet being collected. Also 1,000 tonnes of PVC were recovered and 100 million bottles were re-used.

During the same period, however, waste generation has increased by 3.5% per packaged hectolitre, with an increase of 8% in the quantity of non-refillable beverage containers.

Efforts are thus concentrating on the development of separate collection for the re-use of bottles and on a return to the deposit/refund system.

Germany

Beverage container policy in the Federal Republic of Germany is embodied in the overall waste management policy of the Federal Government. Solutions in that area have, therefore, to be seen in the context of the overall waste reduction and waste recycling policy. This policy was defined by the Federal Government in 1975 in a comprehensive waste management programme. For soft drinks, beer and mineral water containers, the objective was firmly stated: the Federal Government would not allow any further increase of beverage containers in household waste. This left two possibilities for trade (wholesaler and retailer) and industry; either they limit the production and use of one-way beverage containers, or/and they increase the rate of recycled material coming from beverage containers.

This policy was accepted by the industries concerned. In the course of a hearing held in October 1977 in Bonn the following voluntary agreements between government, trade and industry were made:

1. The beverage industry and trade would encourage a more uniform application of the deposit system for returnable containers.
2. The glass industry would reduce considerably the weight of glass containers.
3. The glass industry would increase glass recycling by about 20% per year until 1981; this would bring the quantity of recycled glass to an overall level of about 450,000 tonnes in 1981.
4. The tinplate industry (tinplate producers in the steel industry) would take over all tinplate scrap coming from household wastes.
5. The soft drink industry would not introduce in Germany a high-volume plastic (PET) bottle used as a one-way container.

It was felt that this type of policy, based on voluntary agreement rather than strict regulation, could achieve the environmental goals of the government while leaving some flexibility for the market and also recognising that one-way containers are justified in certain situations on both economic and ecological grounds.

A first assessment of the situation, in early 1979, indicated that the trend towards increased use of one-way containers had stopped. The shares of refillable containers on the German market in 1978 remain rather high, at 97% for mineral water, 94% for beer, and 84–85% for soft drinks. At the same time the recycling of glass and tinplate containers showed a market increase from 1977 to 1978, from 300,000 to 370,000 tonnes per year for glass and from 120,000 to 150,000 tonnes per year for tinplate.

A more thorough study conducted in 1980 by the Federal Ministry of the Interior shows, however, that the market share of refillables has slightly but steadily decreased over the past few years. The main change in this respect was found for soft drinks, with a market share of 81% for refillables in 1979 compared to 87% in 1975. It thus seems that voluntary agreements to stabilise the market share of refillables did not yield the results hoped for. In contrast, the recycling efforts for glass containers and tin-plate packaging were very successful. Glass recycling amounted to 492,000 tonnes per year, and tinplate recycling increased to 180,000 tonnes per year in 1979. As a result, the overall quantity of beverage container-related household waste did not increase during the period concerned.

In an attempt to improve the returnable container system the Federal Government supported a 12-month test with new returnable (glass or plastic) milk containers. The test was completed in June 1981. At the end of the test the new containers had reached a market share of 32% of all pasteurised whole milk offered in 1-litre packages, and the average trippage achieved over the 12 months was 11. The reactions of both the consumers and the participating shopkeepers were encouraging.

Finally, it is worth mentioning that in 1980 a trial took place in the region of Düsseldorf to introduce plastic PET bottles for soft drinks, with a deposit of DM 0.30. It now appears that this type of container will not be further offered on the German market.

The Netherlands

Like most other countries, the Netherlands are experiencing a continuous increase in the use of one-way beverage containers. In 1981, the market share for returnable containers dropped to 96% for beer (compared to 98.5% in 1975), 90% for soft drinks (95% in 1975), and 33% for milk (54% in 1973). Refillable containers are no longer used for wine and spirits.

In 1979 a motion was passed by the Dutch Parliament, requiring the government to develop and propose specific measures (such as mandatory deposits) aimed at the reduction of waste generation from packaging, on the basis of the recently enforced Wastes Act. Apparently, discussions with the industries concerned are still going on.

The most important development in the Netherlands in recent years, however, is in the field of recycling. Following a pilot project launched in 1978 in

North Brabant, involving 71 municipalities and about one million inhabitants, the government used the provisions of the Wastes Act to urge local authorities to organise separate collection of used glass containers. Glass collection skips have now been installed in 80% of Dutch municipalities; the density is about one skip for every 2,000 inhabitants, comparable to the density of postboxes in the country. An average of 35% (i.e. about 10 kg per inhabitant per annum) of the total generation of used glass containers is already recycled in this way, with a number of municipalities achieving more than 50% recovery. It is expected that the level of recovery will soon reach an average of 50%, thus reducing by 5% the total quantity of household waste generated in the Netherlands.

Norway

The system of beverage containers in Norway is based on the following main elements: (i) a return (deposit) system with compulsory standardisation for colour, shape and weight; and (ii) a high tax on non-returnable containers.

For a long time, Norway has had a compulsory deposit-refund system for beer and soft drinks, with a return rate of about 97% and a trippage of at least 22 for beer bottles and 13 for soft drink bottles. A high tax on one-way containers was introduced in 1974. Because of the combined effect of these two elements, the market share of non-returnable containers for beer and soft drinks is less than 0.5%.

In order to prevent this system from deteriorating, or possibly to improve it, other measures have recently been taken by the government. The deposit on returnable containers was increased by 50%, in order to compensate for the past rise in nominal prices so that the balance between the deposit and the cost of a new bottle was re-established. Also the tax on one-way containers, which had remained unchanged at N.Kr. 0.80 since its imposition in 1974, was raised to N.Kr. 1.25 per unit.

Wine and spirits produced or bottled in Norway are also sold in returnable bottles (these beverages are only available through a state monopoly). For some time there was no deposit on these bottles, but the state monopoly was buying the empty bottles back from the customers. At the end of 1978 a deposit-return system was established also for wine and spirit bottles, and the return price was raised to make it equal to that for beer and soft drinks bottles, leading to an increase in return rate from 55% to 75%. These returnable bottles have a market share of about 50%, but this has been declining because of increasing imports of wine and spirits in original bottles, for which there exists no return system. A trend towards more domestic wine and spirits in 1981 resulted, for the first time in many years, in an increase of returnables on the market.

More recent developments concern the market for fruit juices and non-carbonated soft drinks. These are at present sold in non-returnable glass or plastic bottles, or waterproofed paper containers, for which there are no taxes or regulations. During the last few years there has been a tendency for the share of the market held by these beverages to increase, with a concurrent increase in the use of plastic bottles. In addition to the potential litter problem there is a fear that this might lead to a weakening of the present returnable system. The Ministry of Environment intends, therefore, to expand the existing system of taxes and regulations to cover non-carbonated soft drinks.

Sweden

At present in Sweden there is no legislation dealing specifically with waste generation from beverage containers. A tax of 10 öre on beverage containers was introduced in Sweden in 1973, but the purpose was not specifically related to solid waste management. However a private, nationwide deposit-refund system has long been in operation for beer and soft drink glass bottles. The state monopoly for wine and spirits also uses a deposit system for its glass bottles.

Parliament has made statements in favour of recovery, low energy profile for beverage containers and minimal littering, and over the last ten years the matter has been carefully investigated by various research teams and official committees.

As a result of these investigations, a deposit system is to be established for aluminium cans, based on an agreement between the government, the breweries, the trade concerned and the can manufacturer. This system is intended to achieve a recovery rate of at least 75%. It will be run by a specially established 'deposit company', and will start in 1983 to reach its full size in 1985.

Switzerland

Switzerland, like many other countries, has a long tradition of returnable beverage containers through private deposit-refund systems. In 1979/80, taking beer, soft drinks, wine and mineral water together, the market share of refillable glass bottles was still about 85%. However, over the last few years, the trend towards increased use of non-returnable containers has also been noted in Switzerland. For example, between 1970 and 1980, the market share of cans and one-way glass bottles for beer increased from 2% to 12%.

There is no current legislation specific to beverage containers, but the proposed law on the environment contains provisions which enable the authorities to intervene, if deemed necessary. For example, the text under consideration stipulates that the Federal Council may oblige those who sell such products as bottles or mercury cells to take them back, possibly through a refund-deposit system, and that it may prohibit the packaging of consumer goods when such packaging results in excess waste or when it unduly complicates the recycling of materials.

Switzerland is one of the first countries in which recycling of glass containers has been developed on a nationwide, commercial basis by the glass industry in co-operation with local authorities. In general, collection is now organised through large containers with colour separation (for white, brown and green glass), permanently situated in several often-frequented places in about 66% of Swiss towns, representing 84% of the total population of the country. The quantities collected and recycled have steadily increased since the introduction of the system in the early 1970s; in 1980, it amounted to 12.8 kg per inhabitant. These quantities correspond to 35% of the total glass consumption in Switzerland.

United Kingdom

The situation regarding beverage container systems in the United Kingdom has been reviewed by the Waste Management Advisory Council's Packaging and Containers Working Party, a body jointly sponsored by the Departments of the Environment and Industry.

The Working Party examined the energy and raw material requirements of various beverage container systems, and the social and economic implications of any change in the present proportion of the market held by each. It also evaluated the external costs associated with the relevant environmental impacts, including waste disposal, litter and pollution.

The report 'Returnable and Non-returnable Containers', [3] presented to the UK Government in July 1980 after three years' work, shows that the savings resulting from introduction of an 'all-refillable' system in the UK would be very small. It makes the following recommendations:

- Schemes for the recovery and recycling of used beverage containers and other waste should be developed and expanded by collaboration between industry, local authorities, consumers and voluntary organsiations.
- All returnable bottles should bear a clear indication of returnability. For bottles containing ½ litre or more, the Working Party recommends that the symbol should be moulded into all new bottles and that all returnable bottles should carry the symbol on the label or on the closure.
- Manufacturers of beverages in returnable bottles and their retail customers should collaborate to ensure that consumers are informed of the net cost of buying beverages in returnable packs, by a clear indication of the amount of the deposit. This information should be clearly displayed on each individual container or at the point of sale.
- The beverage industries should maintain a continuous campaign to promote bottle returns, as well as an efficient collection system.
- Efforts should continue within the beverage industries to reach agreement on a greater measure of standardisation of returnable bottles.

Discussions have now taken place with the concerned parties, all of whom have confirmed their desire to progress by voluntary means. Progress so far as mainly been made in increasing the recycling of used containers and in introducing a 'returnability symbol' on returnable bottles, as detailed below:

(a) *The 'bottle bank' scheme,* introduced in 1977 by the UK Glass Manufacturers' Federation to recycle *used glass,* is continuing to expand rapidly.

There were 320 cities involved in the scheme by the end of 1981. The glass industry had not quite achieved its target recycling rate of 150,000 tonnes by April 1982, but with a new cullet processing plant about to begin operating near London, it expects to meet the target of 250,000 tonnes per annum by April 1984.

(b) *Can recovery* by can makers and their raw material suppliers is now under way. A new plant for extracting cans from domestic refuse has been commissioned near Manchester and is performing well. Discussions with a number of local authorities for the introduction of 'Can-Banks' are now at an advanced stage and the 'Save-a-Can' scheme for buying back aluminium cans has extended its range of operations.

(c) A scheme for recycling *PET bottles,* which works on the same principle as the glass bottle bank scheme, was introduced during 1981 on an experimental basis at nine sites. First reports are that it is working reasonably well. If the scheme proves to be self-financing at the end of the two-year period, it will be extended to operate on a larger scale.

(d) In addition to the *'returnability symbol'* now being moulded into many new returnable bottles, suitable symbols for inclusion on labels applied to existing returnable bottles are being developed.

CONCLUSIONS

This brief review of government policy in the field of beverage containers shows that the problem is of concern to most OECD countries. Besides previously adopted measures to promote the use of refillable containers or to encourage recycling, the problem is currently under examination by many government committees or working groups. This concern is also reflected in the proposal by the Commission of the European Communities for a Council Directive in this field.

One main reason for this concern is the increasing amount of household waste to be disposed of, and rising disposal costs. This increase is partly due to the continuing trend towards replacement of previously well-established refillable glass bottle systems by various non-returnable container systems. The environmental impact of litter, and the need to reduce consumption of energy and raw materials, also play an important role in governmental concern about beverage containers. Various means are being used to alleviate this problem. They range from the rigid, regulatory approach (such as the ban on all non-returnable containers for the sale of beer and soft drinks in Denmark) to the type of voluntary agreements, involving the different sectors of the economy, now being pursued in the United Kingdom. The approach most generally used in the countries under review is some combination of legislative measures and voluntary agreements.

It is generally claimed that a too rigid and abrupt regulatory approach would create severe disruption in the economy. It does not seem, however, that entirely voluntary agreements have been able to achieve the expected results, with the exception perhaps of glass recycling in those few countries where cullet is really considered by the industry to be an important raw material input to the manufactuing process. This is why, in most countries, the situation is kept constantly under review in order to adjust the approach according to the objectives stated and the results obtained.

Finally, it should be emphasised that, according to the Recommendation of the OECD Council, the primary objective remains the reduction of waste generation through the promotion of refillable containers. Material recycling is another important recommended option, but it represents only the second best solution. In this connection, the achievements of the Swedish deposit-refund system for aluminium cans, once it has been fully established, would be interesting to consider.

Acknowledgements
The views expressed are those of the author and do not necessarily coincide with those of OECD.

REFERENCES

[1] OECD, *Beverage containers – re-use or recycling,* Paris (1978).
[2] Hare, M. J., *Carbonated soft drink packaging in Ontario: An environmental reappraisal (The PET container)* Report prepared for the Ontario Waste Management Advisory Board. August 1979.
[3] Waste Management Advisory Council. *Study of returnable and non-returnable containers,* ISBN 0-11-512939-1 HMSO, London (1981).

The author
Mr Lieben is a Principal Administrator in the Environment Directorate of the Organsiation for Economic Cooperation and development. He is secretary of the OECD Waste Management Policy Group.

CHAPTER 5

Packaging legislation in Denmark

KURT H. GARMIN
Formerly with the Danish Packaging Research Institute, Meterbeun 15, DK2740, Skovlunde, Denmark

---

Packaging legislation in Denmark, with a few exceptions, parallels that in other EEC countries. Danish food labelling rules, in accordance with EEC Directive 79/112, are being introduced and will be fully in force by 1 January 1985. Applications of the e-mark is regulated in accordance with EEC Directive 78/891 and its predecessors.

However, it is possibly of greater interest to focus on those rules which are peculiar to Denmark. The most interesting, from this viewpoint, are the regulations for the marketing of beverage containers. These regulations are based on Act No. 297 of 8 June 1978 on the re-use of paper and beverage containers and the prevention of waste. This act aims to prevent and control pollution by establishing a background for minimising the production of waste and to ensure socially acceptable exploitation, including re-use, of raw materials and products.

The act is limited to paper materials and packages for beverages and milk products. It does not cover transit goods, and normally does not apply to export products, whereas imported goods must comply.
This act authorises the minister of the environment to:

(a) regulate the use of materials in products which would prevent or reduce the re-use of the materials or the products;
(b) lay down rules for the content of secondary and re-usable materials in paper products;
(c) limit or ban the use of certain types of packaging materials and components or proscribe the use of certain materials and types of packaging;
(d) fix rules for deposits on packaging including the sizes of deposits;
(e) place the onus on municipalities to establish procedures for the collection of re-usable materials and products and set rules for such collection arrangements; and
(f) provide rules for control of regulations issued under the provision of the

act including rules for the delivery of information from the manufacturer, the importer and the distributer to the authorities.

As can be seen, this act gives the Minister of the Environment extensive power to regulate and control developments within important sectors of trade, particularly within the field of packaging. To counterbalance this somewhat, and to ensure that not only environmental and resource questions are taken into consideration, the minister is obliged to negotiate with the relevant trade and consumer organisations, and with other ministries affected, before any regulation is introduced. Furthermore, the Minister of the Environment is obliged to inform the parliamentary environmental committee before any important rules are imposed.

On the basis of this act and its more limited predecessor from 1971, the minister issued in 1977 a total ban on one-way packages for carbonated beverages and on 2 July 1981 imposed extensive regulations for the retail packaging for beer and carbonated beverages.

The regulation first of all permits the use only of returnable packaging. It is specifically stated that only packaging which is part of a return system is permitted.

To satisfy this requirement, all packages for the mentioned products must be approved by the state environmental administration on the basis of an evaluation including the following criteria:

(a) a technical evaluation of the package on its suitability to participate in a return system;
(b) whether the return system can ensure that a substantial part of the package after emptying is returned for filling; and
(c) whether an already approved package for the purpose is applicable and available in the proposed volume.

An application must give the following information:

(i) information about type of product for which the package is intended;
(ii) reasons why an already approved package cannot be used and must demonstrate that use of the proposed package does not violate protection of the environment and recycling;
(iii) describe the package, and if possible provide samples;
(iv) provide information on the expected yearly sale in litres and units;
(v) provide a statement which explains how a return system with a deposit will be established, including information on size of the deposit and arrangements for ensuring that a substantial part of the packages are returned for re-filling or material recovery; and
(vi) provide a statement on how regular reports will be made to the authorities on the functioning of the system, especially about the percentage of returned packages and the expected development of sales.

This regulation has been in force since 1 January 1982. Over the preceding 5 years there was a 'voluntary' agreement between the authorities, the breweries and the trade, gradually to reduce the use of beer cans. Since 1 January 1982 it has been permitted only to sell beer in cans manufactured before this time. Carbonated beverage sales in cans have been banned according to earlier legislation since 1977.

One important reason why it has been possible to carry through this regulation without massive opposition is the tradition in Danish beer supply. This tradition is based on two important elements. First, only one officially standardised returnable beer bottle is used and this is easily recognised by consumers. Secondly, the distribution system is managed and, in practice, carried out by the breweries all the way to shops, kiosks, restaurants and canteens. This very strong organisation behind the beer supply system barred for a long time the introduction of one-way containers for beer, and when it appeared, it never reached more than 5% of the total market.

Carbonated soft drinks are mainly manufactured by the breweries and distributed through the beer distribution system. Important producers not affiliated to the breweries are first and foremost Coca-Cola and Seven-Up, but these companies considered it to be in their interest to base their sales entirely on returnable bottles and distribution systems running 100% parallel to the systems built up by the breweries.

In the 1970s some diversification occurred. In addition to the standard 35 cl beer bottle, a new bottle of the same size was introduced by one of the breweries. The 50 cl Eurobottle for beer was also introduced on the market, as well as a 25 cl bottle for beer. Carbonated beverages were marketed in different sizes and a new, generally applied, but not officially standardised, 1-litre bottle was introduced.

At the time the regulations for beer and carbonated beverage packaging were introduced, there were 19 returnable packages on the market, namely:

- 2 officially DS-standardised beer bottles
- 1 DIN-standardised beer bottle
- 2 brewery-owned beer bottles
- 1 officially DS-standardised carbonated beverage bottle
- 1 SIS-standardised carbonated beverage bottle
- 1 carbonated beverage bottle owned by the federation of carbonated beverage manufacturers
- 7 bottles owned by Coca-Cola for Coke, Fanta and multi-purpose
- 2 bottles owned by Seven-Up, and
- 1 bottle for Pepsi-Cola

Some of these were used to only a very limited extent.

These bottles are now approved according to the regulations. How the regulations will be administered in practice is not yet known, but as one aim is to limit the number of types of returnable packages, to rationalise distribution and reduce the costs of sorting and handling bottles within the return system, the application of the rules are expected to be restrictive.

A special problem arising from these regulations is whether they discriminate against imported products. It is obvious that the established Danish distribution system, including the returnable packages now approved, is a rigid system, which favour nearby manufacturers with well-established market positions. On the other hand, it must also be admitted that a number of approved packages are 'open' for use by any manufacturer of beer or carbonated beverages, Danish or foreign, a fact which in many ways favours newcomers, and in certain respects reduces the discrimination due to differences in transport costs.

Closely related, in practice, to the beer and carbonated packaging regulations, is the legislation on package taxes. This legislation covers a much wider range of products, and is concerned with packages for the following products:

- Alcoholic beverages, wine, fruit wine and beer,
- Carbonated beverages and other non-alcoholic beverages and concentrations for such beverages,
- Fruit juice and fruit juice concentrates,
- Vinegar and edible oils, and
- Methylated spirits

For these products, a range of packages are liable to taxes, depending on their size and the raw materials used in their manufacturer. According to the latest increase in the rates introduced in May 1982, the taxes are in Danish øres (US cents in brackets) per package.

Glass and plastics	
10–60 cl	30 (3.6)
60–106 cl	97 (11.7)
over 106 cl	135 (16.2)
Metal	
All sizes	37 (4.5)
Carton and laminates	
10–60 cl	15 (1.8)
over 60 cl	22 (2.6)

These rates are levied on both disposable and returnable packages, but since the tax is paid when the package leaves the manufacturer of the package – for certain carton packages, the manufacturer of the printed material – the tax is charged on returnable packages per filling only by a fraction corresponding to the trippage. For imported, empty or filled packages, the tax is paid when imported. For export products packaged in taxed packages, the exporter can recover the tax.

The differences in the tax rates have caused much discussion and discontent amongst manufacturers of packages burdened with the higher rates, as they feel that this distorts the competition between materials, technologies and manufacturers.

The physical restrictions have obviously been able to direct the use of packages, but the fiscal measures have probably also had some effect. The value of used bottles has increased considerably, and that doubtless led to the increased re-use of akvavit bottles and ordinary wine bottles – types of packages which some years ago were considered as one-way containers despite their technical quality. Furthermore, the introduction of the tax is widely believed to have supported the establishment of a returnable, return-marked wine bottle line now covering an extensive part of the Danish wine supply, particularly for the low-price wines.

In recent years, the Ministry of the Environment has carried out a number of studies covering wine, milk, fruit juice and cocoa milk, to investigate how the introduction of physical regulations on packaging would influence the development, including the application of raw materials, energy, manpower, and foreign currency. The wine and milk areas are not, for the time being, under consideration but limitation in the application of one-way packages for fruit juices and cocoa milk is still an open question.

CHAPTER 6

Product charges in Norway – experiences from the packaging sector

HANS-CHRISTIAN HOLDT
Miljøverndepartementet, Myntgt 2-Dep, Oslo 1, Norway

CHARGES ON PLASTIC BAGS

The production of plastic carrier bags started in Norway in 1965. Grocery stores proved to be important users, as bags were given free to customers buying goods. This brought about a gradual change in shoppers habits; they stopped bringing shopping bags to the stores.

The price of the bags, which at that time was very low, was included in the price of the goods and the customer paid, whether or not he took advantage of the 'free offer'. Plastic bag production had increased to nearly 400 million per year by 1974; equivalent to 100 bags per head of population per year.

Consequently, there was an increase in the dumping of plastic bags. Few studies were made of the extent of dumping, but the increasing number of letters in newspapers, requests to authorities, pressure from environmental organisations, etc, emphasised the problem. This led the Finance Department in the summer of 1973 to start discussing the role of product charges in environmental questions, and a charge on plastic bags was one of many suggestions. The Centre-party government in power at the time warned that taxes on plastic bags and on non-returnable packaging for soft drinks, would be put forward in the coming budget.

This led to a series of articles in the press where commercial interests, plastics producers, and the trade unions stated their opinions on the consequences of these measures.

The Labour party took over after the elections of autumn 1973 and a trade union representative was appointed Minister for the Environment. The Ministry of Environment discussed enforcing product charges internally and decided that they wanted to support these measures as a means of directing environmental policies. The department recognised that the chosen areas were somewhat arbitrary, but accepted them principally as a warning that attitudes were hardening towards less desirable products. It was assumed that these measures would be discussed in advance with the relevant producers, and that problems with employment would not be allowed to occur. The product charge on plastic bags was fixed at N.kr. 0.25 per bag.

The Package Industries Association considered that the use of product charges in environmental politics raised matters of principle, and they requested a complete report on the results before these measures were put into practice. The authorities decided at a meeting in February 1974 that there was no reason to compile a complete packaging report. Due to practical problems occurring for plastic producers, it was decided to postpone putting the product charge into practice from the 1st April until September the same year. However, discussion to limit the measures continued, with the cooperation of the industrial sectors concerned.

For administrative reasons, it was decided that a product charge should be put on the *production and import* of plastic bags. The question then was how to transfer these charges to the user. The Finance Department soon found that they had no authority to order shops to charge for plastic bags.

The Ministry of Environment, in giving their support to the measures, had assumed that the user should pay for the plastic bags, otherwise the product charge was of a pure state income nature and could not be called a 'protection of the environment charge'. Traders pointed out that free bags for their customers gave them an advantage, and many wanted to include these charges in the price of goods, even after product charges were put into practice. The consumer would therefore not be encouraged to reduce the use of plastic bags and the positive environmental results would not appear.

It was claimed that plastic bags were used in the home to pack garbage and that this would stop as soon as bags were charged for. This would lead to considerable difficulties in the disposal of garbage and make the job of refuse collection more difficult. It was alleged that these drawbacks would be far greater than a few bags thrown around the countryside. This problem was seen as a question of teaching, which the schools could take care of to a greater extent.

A debate followed in the spring and summer of 1974 between the producers of plastic and paper bags, as to who had the less damaging product, and a large advertisement for paper bags was reported by the plastic producers who claimed unfair competition. In addition, a series of questions were raised such as 'What can be classified as plastic bag?', 'How should plastic-reinforced paper be treated?', 'What about pre-packed goods in plastic bags?', 'Should large garbage bags in plastic also be covered?', and so on.

The department started to use the following definitions:

- bags which held under 40 litres
- bags with a product price up to N.kr. 0.60
- bags completely or partly of plastic

Bags used for prepacking goods in a factory were excluded.

The absence of a lawful right to charge customers directly for the use of plastic bags, led the environmental authorities to question the product charge matter. The opinion, which had arisen against the system, pointed out that this tax was of a purely fiscal nature and yet another attempt by the government to increase taxes. The Ministry of Environment demanded therefore that a law allowing the retailer to charge the customer directly, should come into existence as soon as possible.

Bags, which were broken down as quickly as, or quicker than, paper were exempt from product charges. The Department received an application for exemption from the plastics producers but the application was turned down as it was not proved that plastics had the necessary qualities.

The charge measures were advertised well in advance, and this led to hoarding of plastic bags. The industry started producing fully in the summer of 1974, and there was a complete stoppage in buying by the 1st September when the measures were put into practice. The means of payment in shops varied. They were not obliged to demand the product charge from the customer, but many chose to do so. Tradesmen reflected on the public's large consumption of bags. The 'oil crisis' of 1973 and the increase in the price of bags, resulted in a considerable expense for shops, and many realised their responsibilities when they saw free bags with their own advertising dumped in the countryside. There was therefore no protest from the retailers' organisation against the introduction of the product charge. There were a few short strike actions in the packaging industry after the introduction on 1st September, to mark their dissatisfaction with the government resolution.

Public reaction was mixed. Generally, the customer reacted less negatively than the plastic industries had expected. The public chose to reduce their consumption when they had to pay for the bags. Those who did not charge for their bags gave them out on demand when the quantity of goods made this necessary. Both the shopkeeper and the public seemed to be more plastic bag conscious regardless of the method being practised.

In December 1974, Parliament took up the question of taxes for 1975. Product charges were also debated and it was resolved that charges on plastic bags should be done away with from 1 January, 1975. The arguments presented in Parliament were much the same as those already put forward by trade interests through the press. The resolution, therefore, was only practised for 3 months, and it is difficult to evaluate the result of these measures. Tradesmen were not allowed to take payment for the bags after 1st January, due to a general price freeze. On the other hand, they received increased profits for other food products to cover the cost of plastic bags. Not until April 1976 were the shops allowed to charge for bags, and they agreed on a price of 20 øre, the same as the buying price.

It soon showed that the consumption of bags had dropped by 25% from 400 million to 300 million. There has been very little increase in consumption since this system was started, and consumption is now at the same level as in 1974.

We can, therefore, conclude that the tax system seemed necessary to interest the grocery sector into taking payment for the bags. The earlier wasteful use of bags has ceased since this system has been in use. Later statistics from the skerries indicate that in spite of an increase in the dumping of plastic materials in general, the percentage of plastic bags in litter has dropped noticeably over the 9 years in which records have been kept. Today's system of payment per bag seems to be an acceptable solution for all parties.

TAXES ON NON-RETURNABLE PACKAGING FOR BEER AND MINERAL WATER

One-way containers in the form of cans for beer and mineral water were introduced into the Norwegian market in 1961. Light pilsner beer was not packed in cans until 1967, and this led to a development in this part of the market. Six years later, 2.6% of beer and 0.5% of mineral water was delivered in cans. The total number of cans was approximately 15 million units.

The relatively rapid increase in turnover led to a new refuse-dumping problem which became increasingly noticeable. The environmental movements which sprang up at the end of the 1960s directed their attention to this problem, and criticised non-returnable packaging from both an environmental and a resource point of view.

Norway had until then been spared the environmental disadvantages attached to beverage containers, as all beer and mineral water had earlier been distributed in returnable bottles. As long as this system was maintained, there seemed no reason to introduce a new system with considerable environmental drawbacks. The attempts to introduce non-returnable bottles for beer in the early 1960s were stopped after a few months after an agreement between the producers. The reason for this was pressure by active environmental groups.

The authorities received a series of enquiries from individual members of the public and organisations who pointed out this problem and demanded that measures should be put into practice to stop the flood of non-returnable packaging. The authorities were aware that increased use of non-returnable packaging would remove the basis for the existing returnable system. Methods for stopping or limiting the use of non-returnable packaging for beer and mineral water were considered against a background of their increasing use. The choice was principally between four measures: prohibition, taxes, quota systems or voluntary limitation on production.

Prohibition of non-returnable packaging was rejected on the grounds that non-returnable packaging in certain circumstances would cause less danger or less bother than returnable packaging. In practice, this was based on a comparison of the damaging effects of tinplate cans and broken bottles and a general prohibition against metal was thought undesirable.

Taxes were regarded as an effective means of reducing the use of non-returnable packaging while still making them available to consumers who, for special reasons, needed such packaging.

The quota system was found to contravene Norwary's agreements with EFTA and EEC.

Voluntary limitations on production would also affect the import of non-returnable packaging in the form of a quota system, and would therefore also conflict with trade agreements.

The tax alternative was, therefore, considered further. The size of the tax caused disagreement between the environmental and the financial authorities. While the Ministry of Environment wanted the tax to correspond to the deposit on returnable bottles (N.kr. 0.50 at the time) the Finance Department thought this was too low to yield the desired results. The tax, based on a rough estimate, was fixed at N.kr. 0.80 when introduced in 1974.

The intention to put a tax on one-way containers was made official a few months before the system was put into practice. This resulted in a wide-ranging debate between the packaging trades and the authorities. It was thought that the resolution would affect the country's only can producer and would have consequences for the 20 jobs associated with can production. The firm, the trade union and the local Labour party were all closely involved in the discussions. They thought that employment should be emphasised and that possible environmental advantages should be secondary to this. Furthermore, a considerable amount of scepticism on the use of taxes in environmental politics was expressed and it was thought that the proposal would not have the desired effect. Furthermore, it was thought that taxes would create a considerable reaction in the general public. The Ministry of Environment in power was from the Labour party and was the union's representative in the government.

In a letter to the Iron and Metal Industries' union, a year after the introduction of the tax system, it was stated that the Department considered reducing the tax from N.kr. 0.80 and that this would be discussed in connection with the national budget for 1976. The matter was twice discussed by the government in the same autumn, but the tax was not lowered. This was not unexpected because, when coupled with the recent withdrawal of the tax on plastic carrier bags, it could have led to political embarrassment.

The system of charges on one-way beverage containers has therefore been maintained, and the fixed charge was raised in the budget for 1982 to compensate for inflation in the period from 1974. The quota of beer cans which fell from 2.6% to 0.3% in 1975 has now risen insignificantly to 0.5%. This corresponds to the share of mineral water, and represents a total annual consumption of 5 million cans.

POSSIBLE FUTURE STEPS

Several conditions are important for a deposit system to work simply. The most important in Norway's case, seems to be the degree of standardisation in the system, both for bottles and crates. The retail link seems to have most grievances against the deposit system, and it is therefore important to simplify the system for the shopkeeper. Possible measures that could be implemented are automatic bottle receivers, standardised bottles and stackable crates, all of which minimise manual bottle handling.

There was a reaction from the retailers and the local bottlers to the introduction of new types of bottles in the 1970s. This led to a voluntary agreement between the breweries in 1978 on type restrictions, where the number of different types of bottles for beer and carbonated beverages was reduced from 19 to 15. Recently, three new ½-litre bottles have been introduced and another two are expected. This has resulted in dissatisfaction on the part of the retailers, and they are now demanding official endorsement of the new bottle types. The authorities have put no restrictions on bottles since the start of the deposit system. It must, however, be considered if some form of official endorsement is to be used to regulate the number of bottle types.

NON-CARBONATED DRINKS

Current deposit regulations have not been applied to non-carbonated drinks. An increasing advertising campaign for such drinks in 0.2-litre non-returnable packaging has, however, stirred interest in the packaging used for this type of beverage. The number of packages today is double the number of beer cans before the product charge in 1974. A decision is to be taken to regulate this kind of beverage.

The quality of bottles has not been subject to official regulation, as the industry itself has set acceptable standards for the production of Norwegian glass. However, an increasing number of bottles are now imported from Eastern Europe, and these are not checked in the same way. There are indications that these imports are more liable to breakage than Norwegian bottles, and the industry should therefore take the initiative to set up rules to check the quality of these imports. The industry is expected to suggest that regulations concerning standards should come under the official resolution for the deposit systems to ensure that they can be legally enforced.

PET plastic bottles have not yet been introduced to the Norwegian market. If these should be introduced, they will today be charged as non-returnable packaging. The authorities, however, are more interested in trying out this kind of packaging in returnable systems. It is therefore natural for PET bottles to be substituted for 1-litre glass bottles, with their weight and safety disadvantages. It is suggested that bottles of 2-litres and higher volume should not be regulated. Such containers are not thought to pose a litter problem.

The author

Hans-Christian Holdt graduated from the University of Oslo with a Masters degree in Economics. He subsequently joined the Norwegian Ministry of Environment as a consultant and in 1980–81 was head of the Recycling Division. Mr Holdt has been a regular representative to the OECD Waste Management Policy Group as well as to NATO CCMS groups on Waste Recycling.

CHAPTER 7

An industrial view of the problems of public intervention in the field of packaging in Finland

JORMA HÄMÄLÄINEN
Finska Forpackinngsforeingen, Ritarikatu 3 bA, SF 00170 Helsinki 17, Finland

---

THE STARTING POINT

In the widest sense, public intervention covers all actions of society in the field of packaging and distribution, using laws and regulations on the orders and recommendations of ministries, central offices and other authorities. In this sense, public intervention would cover, for example, the quality of packaging materials, marking requirements, weights and measures, health, safety, other consumer protection, import restrictions and many other regulations. These regulations are constantly increasing worldwide. In the present context however, public intervention has been taken to mean restrictions introduced to those areas where the determining factors are raw materials, energy, pollution, waste disposal, and recycling of materials and where the product concerned is the one-way/returnable package.

I deliberately want to examine this matter from an industrial viewpoint, and if this paper seems somewhat biased, it is meant to be so. It considers the changes that have occurred in Finland, and, personally, I find it a story about how things should not be handled.

SOME FACTS ABOUT PACKAGING IN FINLAND

The packaging sector has developed in Finland extremely quickly. In the 1950s almost all commonly used products were distributed to retail stores in bulk, i.e. in wooden cases or sacks. Changes started and accelerated decisively in the mid 1960s when industry started to pack their products in consumer packages.

The general conditions which made the change possible were, on the one hand, the creation of a real packaging industry and the enormous development of packaging materials, new applications and solutions, and, on the other hand, the structural change in retailing from service to self-service and from nearby shops to supermarkets.

Today over 90% of regularly used products are industrially packed into consumer packages. Excluding the depression caused by the oil crisis in the mid-70s, the packing industry and the use of packages have grown strongly and noticeably faster than industry in general. In some years during the period 1960–1970, the growth (in value) was almost 50% per year. Today, the packaging industry is one of the biggest industries in Finland. The value of the production of packages exceeded last year FIM 5 milliard (= US $ 1.1 milliard). The share of different materials in packaging are given in Table 7.1.

Table 7.1

Share of different materials used (a) in packaging production and (b) in the use of packaging in Finland in 1982

Materials	Percentage (a)	Percentage (b)
Paper and board	68	45
Plastics	17	23
Metal	10	15
Glass	2	5
Wood	2	7
Other	1	5

In at least two respects, the Finnish packaging industry leads the world. First, almost 50% of the packaging produced is exported; this figure may be unique. The second factor is the use of a highly rationalised distribution system. With the exception of Sweden, Finland probably employs the most advanced distribution thinking in the whole world. The main reasons for this development were the agreement of standardised pallet sizes (1000 x 1200 mm) in the 1950s the adaption of the basic module (600 x 400 mm) for transport packs in the mid-60s, and the adoption of a general module recommendation for the whole of the distribution chain in the mid-70s. These changes were brought about by four big central organisations (basically wholesalers), which account for about 90% of the sales and distribution to the trade in Finland and which had the foresight to develop this distribution system and sufficient power to enforce the other parties to follow.

WHAT HAS HAPPENED IN FINLAND

With the growth of the packaging industry and the general development of consumer packaging in Finland in the mid-60s, it did not take long before both the public and the decision-makers became interested in the matter.

The main cause of this interest was the change in the alcohol laws in Finland in 1968, which allowed the free sale of light beer (under 3.8% alcohol). Before this, the sale of beer was allowed only by the state-owned alcohol stores (at that time about 130 in the whole country). From the beginning of 1969, normal retail stores were allowed to sell beer under a special licence from the state owned company ALKO. Up to that time, beer had been sold in returnable glass bottles, but in 1969, the brewers and retail trade started to develop one-way packages. This action immediately caused discussions aimed at restricting these changes. The main starting points in the discussion were the growing amounts of waste and the need to re-organise waste disposal.

In 1967, there had already been some concern in Finland about the increasing problem of litter. In 1967, the government started a special campaign, 'Keep Finland Tidy', in which the packaging sector was involved through the Packaging Association. The packaging industry also helped financially. The organisers of this campaign made a proposal to the government at the end of 1968, to form a special committee to plan the reorganisation of waste disposal.

At the same time, these problems were also of concern to other Scandinavian countries. In 1968, the Nordic Council discussed a proposal to ban all one-way packages because of their supposed negative influence on the environment. Fortunately, the proposal was not adopted. It was decided only to make a recommendation to the Scandinavian governments to undertake studies to try and find a unified solution to waste disposal. The studies were, as far as I know, never made. After long negotiations, ALKO gave the breweries permission to pack beer in metal cans at the end of 1970. One-way glass bottles were not allowed because, it was claimed, of their even greater littering effects. It is worth noting that the main argument put forward when permitting the use of one-way tinplate cans was that 'ALKO can no longer hinder changes which aim to improve consumer services and to rationalise the trade'. It is remarkable that the one-way package, in those days, was considered a more 'advanced' container than the returnable bottle.

The period of excitment continued. The committee for waste disposal issued a part-report at the beginning of 1970 supporting restrictions on the use of one-way packages. The matter was immediately discussed in parliament. On 15 April 1970 the government decided that one-way glass bottles for carbonated non-alcoholic beverages should be banned and the introduction of beer cans should be postponed.

The packaging sector reacted strongly, emphasising that the proposed prohibitions did not solve litter problems, but that a waste disposal organisation should be created and the task of collecting information should be increased.

At the same time, it was complained that the packaging sector had not been invited to participate in the waste disposal committee, that the representatives of the sector had not been heard at all, that the Packaging Association had not been asked to comment on the report and that a copy of the report had not even been sent to the Association even though it had requested one.

So in the short history of packaging in Finland, discussion of this matter had clearly become inflamed. Even in 1970, it was possible to pick up a ready-written government proposal to parliament banning all one-way packages for beer as well as other kinds of carbonated beverages. This proposal was, however, never brought to the parliament. The packaging branch fought for its life, and instituted a wide programme of studies and information.

The committee for waste disposal issued its final report at the beginning of 1971. The main content of the report was a proposal to create a nationwide organisation for waste disposal. At that time, individual householders were responsible for their own waste disposal, and private companies took care of the transportation. It was also proposed that the government should be given certain rights to ban or restrict the use of packages that were thought likely to cause littering.

However, the pressure for the introduction of new packages was great and so, at the end of 1969, ALKO gave the breweries permission to start marketing beer in tinplate cans in March 1971. ALKO, however, set the price for canned beer somewhat higher (30–35%) than for beer sold in returnable bottles. The share of beer sold in cans was estimated to be about 5%.

Then, after some years of hard fighting, the situation calmed down. In 1972 The Packaging Association organised a committee to investigate the re-use of packaging waste. Public discussion became more sensible; at the same time, it noticeably decreased. In 1973, a special tax was introduced in Sweden. In Finland, however, no special tax was implemented. In 1973, the further discussions were held on the development of the returnable system. In 1974, both energy and recycling problems were discussed within the Packaging Association.

At the end of 1975, things started to move again. In the 1976 state budget, the government introduced into parliament a proposal, according to which a new tax (in addition to the other already existing excise taxes) would be levied for beer (36 pennies/litre) and other carbonated beverages (60 pennies/litre). It was justified as a measure to protect the environment, especially against littering. Once again, it must be noticed that the packaging sector was not consulted in advance nor was parliament interested in listening to those with some expertise. The tax then became valid from the beginning of 1976.

The tax for beer of 36 pennies/litre is still valid today. At the beginning of the 1970s when canned beer was first marketed, ALKO let it be known that they wanted to ensure that canned beer would not exceed a certain limit. This limit soon turned out to be 10% of the total beer sales. Having the exclusive right to fix the prices for beer, ALKO has kept the share of canned beer to under 10%

of total sales by always setting the price of canned beer higher than the price of beer in returnable bottles. Today, it is 38–55% higher. In the beginning, ALKO said that the revenue resulting from the price differences would be used for environmental purposes. This, however, has never been done.

Along with the tax laws, the government in 1975 presented to parliament a proposal for a waste disposal law based on the previous committee reports. For the packaging sector there were many positive points in the proposal, for example, a general littering prohibition and an obligation for communities to organise waste disposal in their own area. One serious defect of the proposals from the viewpoint of the packaging sector was paragraph 34 of the proposal, according to which, the government would be given an open power of attorney to restrict the use of a certain type of package, or even ban it or set a special tax or returning charge on it, if it was shown that the package caused significant extra costs or difficulties for waste disposal.

The packaging sector strongly resisted the proposal, which was thought to give the government powers that were too far-reaching. Nevertheless, the law including paragraph 34 was ratified and came into force at the beginning of 1979. I still think that the powers are too wide, particularly as the opinions of the industry and trade are neither sought nor wanted when new actions are being drafted. Thus, it is too easy to be driven into over-hasty and wrong decisions.

In the following years, the situation remained unchanged, as did the amounts of the taxes. The use of one-way packages did not increase significantly. At the end of 1980, the tax on carbonated beverages was increased to 1 mark/litre in line with inflation. The increase had no noticeable effects.

At the beginning of 1981, matters took a new turn. The Department of Environmental Protection in the Ministry for Internal Affairs proposed to the government that one-way packages for carbonated beverages (beer excluded) should be completely banned. This action was based on paragraph 34 of the waste disposal law. The arguments put forward were:

(a) the use of carbonated beverages in one-way packs had increased to 1.9% of the total sale of those beverages in 1980, being 1.2% in 1979 and 0.7% in 1976;
(b) the number of one-way packages was correspondingly 11 million in 1980, 6.6 million in 1979 and 4.0 million in 1976;
(c) the main part (85%), consisted of imported products such as Coca Cola and Pepsi Cola which were packed abroad, and
(d) the development threatens to destroy the returnable glass bottles system which worked well.

The whole of the packaging industry and the trade strongly opposed the proposal, which was then quickly and quietly withdrawn.

Only a few months later (June 1981), the Ministry for Internal Affairs proposed to the Ministry of Finance that the package tax for carbonated beverages in one-way containers should be raised from 1 mark to 4 marks/litre. The arguments were as before but with the addition that the extra money collected by the higher tax would be used (so they claimed) to support the waste disposal in the Archipelago and Fjeld districts – what a wonderful decoy for the politicians!

The consequence was a new fight and, at the request of the industry, even a consultation between the authorities and representatives of the packaging sector. The result was an increase in the tax for one-way packages made of glass or metal to 3 marks/litre while for those made from other materials would be taxed at 1 mark/litre as earlier.

LATEST DEVELOPMENT

In May 1982, one of the leading breweries in Finland announced that they were to start the sale of some so-called refreshing drinks (carbonated) in the Swedish made Rigello one-way bottles (Barex/board) and another big brewery publicised the start of test-marketing Coca-Cola in 1.5 litre PET-bottles made in the Netherlands. For the time being, both are to be packed abroad (Sweden/Holland), but according to both companies, it is planned to move the packing to Finland, if sales look promising.

THE FUTURE

The Finnish government has recently proposed that parliament should set up a Ministry for Environmental Matters, in which all matters concerning environment and natural resources would be concentrated. In principle, the proposal is good. However, it could be bad for the packaging sector if the officials are chosen on political grounds and are forced, especially during the first years, to prove the necessity of the ministry and their own posts and, what is worse, their effectiveness to their own parties by developing an enormous number of initiatives and proposals. Many such proposals are likely to be completely useless and impossible to implement but will cause a great deal of confusion and uncertainty about the future.

Today, Finland still lacks an independent packaging research institute such as Packforsk in Sweden or PIRA in England. To remedy this, the packaging sector – about forty companies representing all parties interested in the packaging field, together with the Technical Reseach Centre of the State – has, after preparation for over two years, just started a 'Research Project on Packaging Technology' located in the STRC. The project will last for 5 years and the budget is FIM 600,000 (= US $135,000) per year. The packaging sector is to finance half of the project, the STRC the other half. The project aims to support research and product development in the entire field of packaging and distribution, mainly through special projects and thus promote and develop activity within the sector. One special aim is to initiate university level education in packaging and packaging techniques. It is also expected that the project will achieve the position of an independent authority, on whose opinions decisions in the future can be based.

CONCLUSIONS

I do not wish to comment on the situation in other countries, still less to give advice to anybody, because the circumstances differ from one country to another so that the solutions must be tailored to local needs. As far as Finland is concerned, I only hope that in future, solutions are based more clearly on practical information than on pre-conceived ideas and prejudices. When considering different actions in the packaging field, it is vital that the effects of all the various alternatives should be weighed together.

It is, however, often very difficult to compare various alternatives because, in addition to costs and other unambiguously measurable values, attention must be paid to abstract values, whose significance invariably depends on the personal experience of the person making the comparison. When considering abstract and other values not directly measurable in economic terms, the main question is, I think, how much we are willing to pay for certain real or imagined advantages. One thing, however, is certain. Everything must eventually be paid by the consumers, only they are never asked in advance.

Acknowledgement
The views expressed in this paper are those of the author and do not necessarily coincide with those of his parent organisation.

The author
Jorma Hämäläinen is Managing Director of the Finnish Packaging Association in Helsinki.

CHAPTER 8

Present and impending regulations concerning packaging and the environment in the Netherlands

A. Th. van EWIJK
Ahold N. V., Postbus 33, 1500 EZ Zaandam, The Netherlands

More than 10 years ago, on 10 October 1971 the Opening Report of the Expert Group on Packaging of UNIDO (United Nations Industrial Development Organisation) stated that packaging should be considered part of the industrial infrastructure. It is easy to predict the calamities that would ensue if a few well-meaning civil servants toured the land destroying half a mile of tarmac, breaking a few railway switches, randomly cutting telephone lines or dynamiting a bridge or two. Such actions would be termed sabotage. Legal measures interfering with complicated packaging systems could have similar, if a trifle less calamitous, consequences. Nevertheless Dutch authorities have shown, and continue to show, a certain eagerness to interfere legally with packaging.

It is well known that government interference with packaging issues is one of the less pleasant aspects of the packaging industry. There is no doubt that the use of non-refillable containers generates waste, and too much waste is detrimental to the natural environment. This environmental impact of waste from packaging is a cause for concern. Over a decade has passed since the MPs Epema-Brugman and Terlow (the present Minister of Economic Affairs), first expressed their worries on the subject and proposed legal measure against packaging in order to reduce the annual flow of waste. Similar worries in the business world led to the establishment of the Foundation for Packaging and the Environment. The Foundation was meant to add the industry and trade's view to the thinking on problems in the Packaging and Environment field. This contribution consists primarily in the promotion of a policy that envelopes effectiveness, marketing, overall efficiency and other aspects of management, in equal consideration with environmental impact and social costs and benefits. A policy, in other words, that serves both interests; environment and packaging. Environmental impact, in the Foundation's view, should not be taken as either a minor or a dominant factor, but precisely equivalent to other factors.

The Afvalstoffenwet (Refuse Law) of 23 June 1977, is the centrepiece of government policy regarding Packaging and Environment. Chapter 5 of the law, entitled 'Protection of the environment by the reduction of waste' opens possibilities for restricting measures against packaging. The Minister of Health and Environment is entitled by article 27 of the law to ban any goods (the manufacture and sale thereof) which, after having been used for the purpose they were made for:

(a) are to a less than desirable extent, re-usable or adaptable to re-use;
(b) are difficult to dispose of effectively, otherwise than through re-use;
(c) add unduly to the amount of waste; and
(d) enter the environment as litter to a considerable degree.

It follows that it should not be too much of a problem for the Minister to ban non-refillable containers because:

(i) non-refillable containers can obviously not be used twice for the same purpose;
(ii) as long as recycling is still comparatively undeveloped, the disposal of the waste arising from this type of packaging does indeed pose problems;
(iii) who decides what should be termed as 'undue addition to waste'?; and
(iv) a proportion of litter does indeed consist of used non-refillable containers.

Article 28 contains an even clearer statement:

> In the interest of promoting re-use or any other means of limiting the production of waste with the purpose of environmental protection, regulations may be brought into effect concerning the use of packaging or packaging materials and goods produced to be used once.

This article also provides the legal basis for mandatory deposits. Chapter 8 of the Law provides for the imposition of taxes and duties. None of the regulations for which the Refuse Law provides the legal basis has as yet come into effect. This is mainly due to the fact that the former Minister of Health and Environment proved receptive to the argument that the far-reaching consequences of such regulations justified a thorough consideration of alternative ways to reach the goal of waste-reduction. Talks between government, industry and trade were held under a gentleman's agreement. The parties agreed that industry and trade would voluntarily contribute to reducing refuse; the establishment of a bottle bank network, other ways of separation of waste material at the source (cans, paper) and efforts to improve recycling are such contributions.

The Refuse Law, however, is not likely to remain the only law which enables the Minister to restrict packaging. It is probable that proposals on a new Food and Drugs Law will be submitted to Parliament. The proposals contain an environment section. At present, the precise text of the section has not been published, but a Health and Environment spokesman has stated that the new environment section provides for restrictions and for banning of products which, when used by the consumer, may directly or indirectly affect the environment! Packaging is to be one of the products the Law aims at.

Two other elements in the present proposals also give rise to concern:

(a) The definition of 'goods' in the new Food and Drugs Law has been widened to such an extent that regulation on the basis of this Law can affect even early stages of production.
(b) Article 14 of the proposals enables the minister to impose a restriction or a ban by a mere decree. Admittedly a restriction or ban would only be temporary (one year maximum), but the practical consequences of a one-year ban may amount to the same as those of an indefinite ban.

Two new legal projects in the environment field deserve mention. A proposal was made in May 1981 to introduce the Dutch equivalent of the Environmental Impact Statement. The new law has not yet been worked out in detail so it is not clear precisely what sort of activities will be subject to Environmental Impact Statements. For example, a company about to market a new product or introduce a new packaging system or initiate any other activity, might in future be required to state:

(a) the goal it seeks to achieve with its new activity;
(b) alternatives to its activity or product;
(c) the present environmental situation and its expected future development if the company's activity is not introduced, nor any of its alternatives;
(d) the expected environmental impact of the new product or activity;
(e) a comparison between (c) and (d).

Authorities will then decide on the desirability of the new product or activity. Though the scheme is still *in statu nascendi,* and its field of impact as yet not well known, it deserves attention.

The second project worth noting is the move to extend the existing 'Wet Algemene Bepalingen Milieuhygiëne' (General Environment Regulations Law) into what is termed a 'basic' law meant to integrate 'Environment' into overall government policy.

These then are the main provisions and possibilities for legal measures on packaging and environment in the Netherlands. The extent to which these measures will eventually come into effect depends of course on the intentions of the responsible Minister, and his or her political muscle, as well as the mood in Parliament on the subject. In 1979, a Parliamentary resolution stated the following order of priorities for what were termed the 'various stages of waste-prevention and waste-disposal':

(a) a *preventive* policy with regard to waste;

(b) re-use of raw materials from waste through separation of the waste material at the source;
(c) re-use of raw materials from waste involving mechanical schemes for the processing and separation of bulk-waste;
(d) conversion of suitable waste into energy;
(e) controlled dumping of non-recycleable waste or residue;
(f) other means to realise the above mentioned objectives, including research and development.

More recently, Parliament called for a swifter implementation of Chapter 5 of the Refuse Law. These two parliamentary resolutions seem to indicate that the legislators are in the right mood to support a zealous Environment Minister. Unfortunately, it seems that at the moment we have zealous Environment Minister with definite intentions to deal with household waste, 25% of which is packaging waste. The Minister does not favour burning household waste, not even as RDF (refuse derived fuel) or agreements with trade and industry on voluntary waste reduction. This latter topic deserves some amplification.

Mention has already been made of the talks that were held with the previous Minister. Partners in the talks were commodity boards and business organisations and the Minister of Health and Environment with the Foundation for Packaging and Environment acting as intermediary and coordinator. The former Minister held the view that the relation between packaging and environment, and the ensuing problems, were of a dynamic nature calling for flexible solutions. He considered agreements to offer the best opportunties for such flexibility.

The present Minister however, holds a different view. Whatever agreements have been, or will be, made, Chapter 5 will be implemented. She (the Minister) also tends to involve consumer organisations and environmentalists in the new round of talks, and has explicitly stated her preference for the introduction of a so-called IMP on waste. IMP stands for 'Indicatief Meerjaren Programma' which is in essence a sort of Five Year Plan on Waste. The IMP will include several measures implementing Chapter 5, such as mandatory collection of glass bottles and jars by means of bottle banks and increasing the use of refillable containers. Mandatory deposits are included in the ideas. The Minister intends to align the proposals with plans existing in the Federal Republic of Germany to impose a levy on non-refillable containers. This IMP will be submitted to Parliament in the Autumn of 1982 and the Minister hopes to start its implementation during 1983.

In order to end this short overview of present and impending regulations on packaging on a lighter note, I would like to quote our Secretary of Economic Affairs, whose department has some influence on the implementation of the Refuse Law. He has said 'I don't think it has yet been established to what extent non-refillable containers are damaging to the natural environment'. The Foundation for Packaging and the Environment has succeeded in promoting the idea of the importance of packaging in economic processes and the complexity of packaging systems. As a result, it is expected that the Department of Economic Affairs may curb any over-zealous environmental regulations restricting packaging that emanate from the Health and Environment Department.

CHAPTER 9

Limitation of energy consumption and domestic waste production in the field of beverages

JEAN-CLAUDE COHADON
Peat, Marwick Mitchell and Co., Tour Fiat, Cedex 16, 92084 Paris la Defense, France

INTRODUCTION

Reducing energy consumption and domestic waste production in beverage packaging and distribution is currently an issue of topical interest. This chapter therefore presents the results of experiments conducted in France over the last few years. In outline, the chapter gives a brief résumé of the background to an agreement signed by government and industry sectors, the terms of this agreement and the characteristics of the system used to monitor it. The final part of the chapter provides an overall picture of the French beverage industry, the objectives which should be achieved by 1984 and the results of the first two years of the agreement. The chapter concludes with a brief discussion of the change in energy consumption between 1981 and 1976 (the reference year).

BACKGROUND TO THE FRENCH 1979 GOVERNMENT – INDUSTRY AGREEMENT

Since 1974, sharp increases in the prices of energy and, more generally, of raw materials, coupled with the increasing consumption of these resources in France, led the French government to edict and follow a policy to reduce energy consumption. Two official bodies were created to promote this policy: L'Agence pour les Economies d'Energie (Agency for the Reduction of Energy Consumption) created by a government decree dated 27 November 1974, and La Délégation aux Economies de Matières Premières (The Delegation for the Reduction of Raw Materials Consumption) created by a government decree dated 23 April 1976. The purpose of these agencies was to develop objectives and promote action plans for saving energy and raw materials.

As early as 1974, the possibility of enforcing the use of returnable containers, mainly bottles, was seriously considered, and a study of the economic impact of such a measure was made at the request of key industry sectors involved, namely, manufacturers of glass, plastic and board containers, bottlers, wholesalers and retailers' organisations.

Two prime concerns of the study were to evaluate:

(a) the energy required to manufacture the necessary stock of returnable bottles and crates compared to the annual energy consumption required by the existing system of returnable and disposable containers – the initial investment could only be recovered after six to eight years according to various hypotheses;
(b) the investment costs for bottlers and transporters, as well as disinvestment costs for container manufacturers (glass makers, plastic industries and paper and board industry).

In the light of the results obtained from this study, the idea of enforcing a full system of returnable containers was rejected because of its major economic implications for all sectors involved. However, this study, fully endorsed only by the economic sectors concerned, enabled them to present the government with a set of measures which could be adopted by industry in order to reduce energy consumption in the bottling and distributions of beverages. At that time an agreement was reached between government and industry.

Two years later, in 1977, industry and government decided to launch a joint study for evaluating different scenarios for energy and cost savings. Scenario, in this sense, was defined as a hypothetical situation, 'What would be the result if a change in containers or outer packaging manufacturing were requested?'. More than fifty scenarios were developed and studied by government and industry. All scenarios were evaluated in terms of investment and disinvestment costs, implications for professional sectors, increases in consumer prices, changes for the consumer, effect on employment, influence on the balance of payments and so on.

This was the first joint effort by government and industry to devise a solution for energy savings in this area, and the study necessitated development of a model which was then computerised, and served as a mathematical model for all the evaluations. Increased impetus was given to the work by a growing general concern regarding waste and pollution as well as environmental protection. The beverage industry was particularly affected by this concern about energy consumption and environmental pollution, on account of rapidly developing disposable containers. It resulted in the creation of the Agence National pour la Récuperation et l'Elimination des Déchets (ANRED), (The National Agency for the Recovery and Elimination of Waste Materials).

In 1979, the French Government was obliged to take some action, and decided to launch a series of discussions with industry to find a solution. The following industrial sectors were consulted:

(a) container manufacturers (glass, plastic, cardboard);
(b) manufacturers of outer packaging (crates, boxes, packages, cartons, etc.);
(c) bottlers;
(d) transporters;
(e) wholesalers and retailers;
(f) organisations involved in the recovery and elimination of waste materials.

After several months of discussion, an agreement was signed on 17 December 1979 between the government, represented by the Ministry of Industry and the Ministry of Environmnet and Quality of Life, and the professional sectors from the beverage packaging, bottling and distribution sectors. Other Ministries were, however, involved to a greater or lesser extent: the Ministry of Agriculture – in connection with problems related to wine and milk; the Ministry of Economy and Finance – as regards price variations for consumer goods; and the Ministry of the Interior – as communities were directly concerned by certain measures recommended in selective waste collection. It should be noted that during this same period, a draft bill requesting a tax on disposable containers was presented by a member of the Socialist Party. This bill was not accepted by the French Parliament.

THE 1979 AGREEMENT

The agreement signed by the parties in 1979 covered a five-year period with objectives being set for 1984. Of particular interest among the principal clauses of this agreement are:

(a) A reduction of 12% over a five-year period of the energy consumed in the bottling, packaging, and distribution of beverages.
(b) Eight beverages were included in this agreement: wines (other than sparkling), mineral and spring waters, milk, beer (including shandies), soft drinks (carbonated drinks, colas, bitters, tonic waters, fruit drinks and assimilated beverages), edible oil, alcoholic beverages and fruit juices. The scope of this study excluded sparkling wines such as champagne, ciders (apple or pear beverages) and vinegar.
(c) A reduction of 40% over the same five-year period of domestic waste related directly to the disposal of beverage containers.

A general survey of the domestic consumption of beverages had already been conducted in connection with the earlier 'scenarios' study. Since the public authorities had no other basis for comparison, these 1976 results were selected as the reference point and the savings indicated above were related to this reference level.

The means of achieving the above objectives remained the responsibility of the professional sectors, by considering:

(a) technical innovations;
(b) reductions in the weight of containers;
(c) development of means for re-using bottles, either by compulsory deposits or use after selective collection;
(d) recycling of used materials (an objective of recycling 600,000 tons of cullet by 1984 had been proposed); and
(e) regeneration of plastic materials, etc.

This agreement was to cover only the consumption of beverages bottled and consumed in France, thus excluding imported products bottled outside the national territory. In addition, the agreement signed between the public authorities and industry, required the setting up of an instrument to measure the annual progress towards the 1984 objectives. This instrument of measurement – or 'barometer' – was developed by Peat, Marwick, Mitchell & Co. Consultants (France) and was attached as an appendix to the text of the agreement signed on 17 December 1979. This measurement 'barometer' was directly derived from the calculation tool used for the earlier 'scenarios' study.

THE BAROMETER

Ideally the 'barometer' should be a means of identifying and measuring changes that would be acceptable to all the parties concerned. However, it was recognised that it could not take into account *all* the economic and sociological phenomena inherent in the packaging of beverages, such as air pollution, employment, balance of payment consequences, consumers attitudes, etc. So finally it was agreed that the methodology to be adopted would focus mainly on two major parameters:

(a) the consumption of energy (the measurement unit to be adopted being the quantity of energy per hectolitre bottled); and
(b) the production of waste (the unit to be adopted being the quantity (in kg) of waste per hectolitre bottled).

The end product of this exercise is an annual report, submitted each summer, and dealing with:

(i) changes in the consumption of each of the eight beverages, as well as changes in their packaging (types of containers) or distribution patterns;
(ii) calculation of two indices; the energy consumption (kg of oil equivalent per hectolitre of beverages) and the domestic waste production (related to the consumption of beverages bottled and consumed in France and expressed in kg of waste per hectolitre);

(iii) an analysis of the reasons for any changes in the indices according to evolution of market demand, weight reduction in containers, utilization of waste (e.g. re-employment of used bottles, recycling of glass), improvements in glass industry technology and improvements in plastics industry technology.

Many difficulties were initially encountered in the execution of this project, for both technical and psychological reasons. The methodology used today is the result of numerous and extensive studies undertaken since 1974 for both the public authorities and private enterprises in the economic sectors concerned. As a result of this, the technical knowledge gained, as well as, perhaps more significantly, the confidence established with the different participants, since all projects undertaken were done so in the strictest confidence, have been largely responsible for the success of the mission and the acceptance of the Consultants by all the parties concerned.

From a technical point of view, the most difficult concept to develop and establish was the cost concept, which had to be:

(a) homogeneous for all products and companies;
(b) meaningful in that there should only be a slight discrepancy between similar companies in the same economic sector;
(c) reliable and measurable;
(d) politically and economically acceptable, as the confidentiality of certain economic and financial information submitted by companies should be respected, especially concerning profit margins.

In addition, the cost concept adopted had to:

(e) evaluate the consequences of any sizeable variations in, for example, the volume sold, transfer of the market from one distribution channel to another (a decrease in sales of a given product offered in hypermarkets to to benefit of small, local grocery shops, etc.), a reduction in the weight of some containers, a modification in the use of packaging material (for instance, the shift from cartons to polyethylene films, etc.);
(f) produce objective and reliable results, without the need for further elaboration.

In order to take account of all these constraints, it was thought best to use the notion of direct costing as illustrated in Fig. 9.1.

To carry out an annual run of the model, basic data are gathered from several companies, government and professional bodies. Each piece of information collected from one operator is verified against that collected from another. For instance, glass makers' data, such as the weight of the bottles, are checked against the information provided by bottlers, who are their clients.

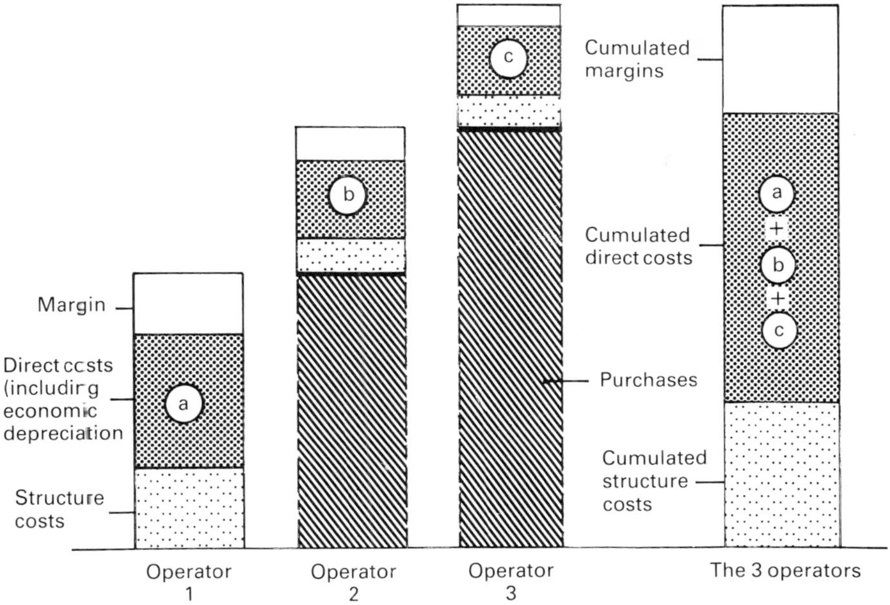

Fig. 9.1 – Schematic diagram illustrating the method of direct costs arising as a result of the successive operations by three operators. Only measurable units, such as therms of energy, minutes of labour time and francs of investment are used in the model.

Processing of the model is linked to three basic phenomena:

(a) Sales of beverages for each type of container and for each distribution channel. Some 70 different types of containers are included in the 1981 'Barometer'. For instance, 11 types of containers are retained for beer; namely 2 types of barrels (50 litres and 30 litres), 5 types of returnable glass bottles (100 centilitres, 75 cl, 65 cl, 33 cl and 25 cl), 3 types of disposable glass bottles (75 cl, 33 cl and 25 cl), and 1 miscellaneous (mainly cans which account for a very small percentage of beer sales in France).

(b) Distribution channels. These cover cafés, hotels, restaurants and 'collectivities', as well as food store circuits, broken down into hypermarkets (self-service stores with a sales area of more than 4,000 sq.m), supermarkets (self-service stores with a sales area of less than 4,000 sq.m), and traditional and local grocery stores.

(c) A description of containers and outer-packaging use for distribution. By way of example, the sales of mineral waters in PVC bottles of 150 cl are packaged as:

   (i) 35% of loading trays are made up of 50 cartons (weighing 375 grams) each holding 12 containers (46.6 grams);

(ii) 5% of loading trays are made up of 100 packs (30 grams of polyethylene) each holding 6 containers; there are 6 layers of packs separated by 5 dividers (950 grams of carton) and covered by a sheet of polyethylene weighing 1,275 grams etc.

Costs are evaluated in terms of energy, labour and economic depreciation, for all operations from extraction of raw materials required in a container's production or its importation to its disposal or recycling. The 1980 Barometer used some 225 basic operations:

(a) 9 operations for materials processing (glass, PVC and polyethylene . . .);
(b) 36 operations for container manufacturing (small and large glass bottles, various types of extruding machines for PVC bottles . . .);
(c) 82 bottling and outer-packaging operations (according to the speed of the bottling machines, washing, rinsing, etc);
(d) 8 operations for handling and transportation;
(e) 16 operations for wholesale distribution;
(f) 50 operations for retail distribution;
(g) 24 operations for disposal and recycling of material (various types of waste collection, selective collection of glass or PVC bottles, various treatments of waste products etc.).

In addition the 'Barometer' is a 'living' tool and can be readily modified to take account of new products, such as 'panche' (shandy), new materials, such as 'protected glass' and new containers, such as 'briques' for fruit juices. These can be incorporated when they become significant in the market. The selection of containers used in the Barometer is revised annually.

The sampling of companies providing cost data for the various sectors is wholly representative. However, owing to a greater or lesser degree of integration of the industries concerned, sampling percentages range from 20% of the wine industry and 30% of the milk industry to 90% or more of the industry for beer, mineral waters and edible oil.

THE FRENCH BEVERAGE INDUSTRY

Table 9.1 shows the total volumes of different beverages bottled and consumed in France in 1981. Note that these exclude sparkling wines such as champagne, apple and pear ciders, vinegar and syrups.

The proportion of beverages imported or exported after bottling is insignificant. Bottled imports of the beverages concerned represent only 3.7 million hectolitres (mainly beer, milk and alcoholic beverages). Bottled exports amount to 10.4 million hectolitres (mainly wine, mineral waters, alcoholic beverages).

French consumption of beverages stands at 152.6 million hectolitres, 3.7 million hectolitres of which are imported; thus the French consumption of beverages bottled and consumed in France amounts to 148.9 million hectolitres.

Table 9.1

Production and comsumption of home-bottled beverages in France in 1981

Beverage	Bottled in France		Bottled *and* consumed in France
	10^6 hectolitre	% of total	10^6 hectolitre
Wine	46.0	28.9	42.885
Mineral & spring water	34.6	21.7	30.580
Milk	31.7	19.9	30.990
Beer	22.2	13.9	21.174
Soft drinks[a]	12.7	8.1	12.532
Edible oil	5.1	3.2	4.514
Alcoholic beverages	4.7	2.9	3.236
Fruit juices	1.8	1.1	1.620
Panache (shandy)	0.5	0.3	0.517
Totals	159.3	100.0	148.048

[a] Includes carbonated beverages, lemonade, colas, tonic water, bitter lemon, fruit beverages, etc.

Of this, it has been estimated that 0.9 million hectolitres are bottled in unusual containers which are not surveyed in the Barometer. The reference figure for the 1981 Barometer is 148 million hectolitres (i.e. 99.4% of the volume of beverages bottled and consumed in France).

Wine

Total consumption of bottled still wine has tended to decrease (42.9 million hectolitres in 1980/81 as compared with 46.5 million hectolitres five years before), as shown in Fig. 9.2. By considering the type of containers used, Fig. 9.2 shows that following a sharp decrease between 1975/6 and 1979/80, sales of returnable glass bottles are stabilising and sales of wine in non-returnable bottles should now plateau after a substantial increase in previous years. PVC containers are becoming used less and this trend is expected to continue, although 5–20 litre containers (e.g. polyethylene bags) should continue to develop.

Mineral and spring water

As shown in Fig. 9.3, there was a significant decrease in total consumption between 1976 and 1980, although 1976 was an exceptional year because of the

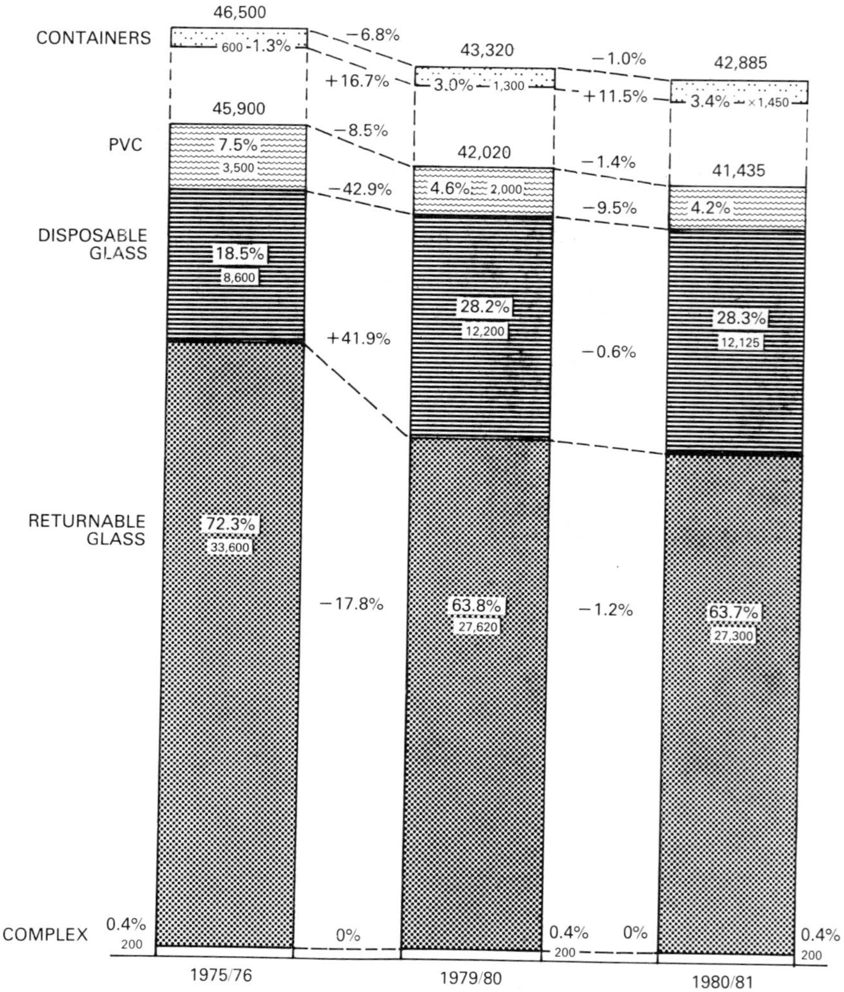

Fig. 9.2 – Pattern of wine consumption in France for 1976, 1980 and 1981. The data refer to still wine only and exclude direct consumption by producers. (Data based on industry sources.)

unusually hot summer. In general, sales of mineral and spring waters are slowly increasing (30.6 million hectolitres in 1981 compared to 30.1 million hectolitres in 1980). About 85% of all sales are in PVC maxi-bottles, but sales in returnable glass bottles are slowly decreasing; they represent only 11% of sales as compared with 18% in 1976 and are to be found mostly in restaurants and bars. Sales in non-returnable glass bottles are increasing slightly but the volume sold in this type of container remains low at less than 4% of total sales.

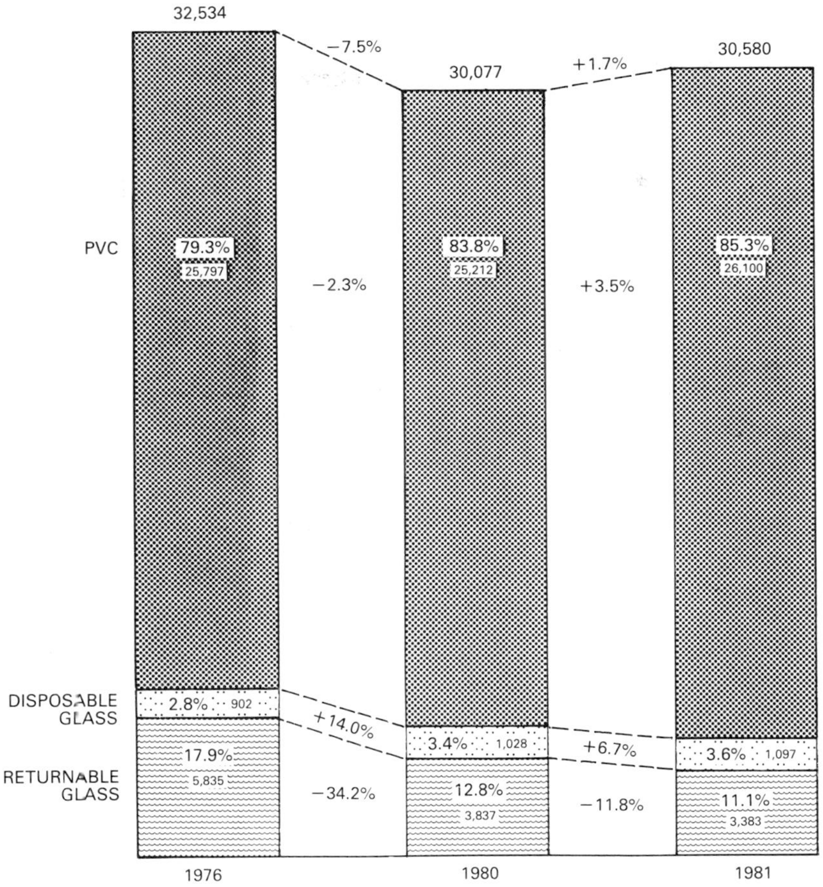

Fig. 9 3 – Pattern of consumption of mineral and spring waters for 1976, 1980 and 1981. (Data based on industry sources and Syndicat des Eaux Minerales.)

Milk

The consumption of milk shown in Fig. 9.4 seems to indicate a significant increase between 1976 and 1980/81. However, this change arises from the statistical methods used to evaluate sales and does not have any practical effect on the figures derived from the Barometer.

In France, milk consumption is very stable and increases slightly with the population. Annual sales amount to 31 million hectolitres. All milk sales are in non-returnable containers, except for a small proportion (2.5% of total sales) delivered to cooperatives in 20-litre containers. Sales in 'bricks', cardboard, polyethylene and now aluminium containers are expanding. Polyethylene bottles are used less.

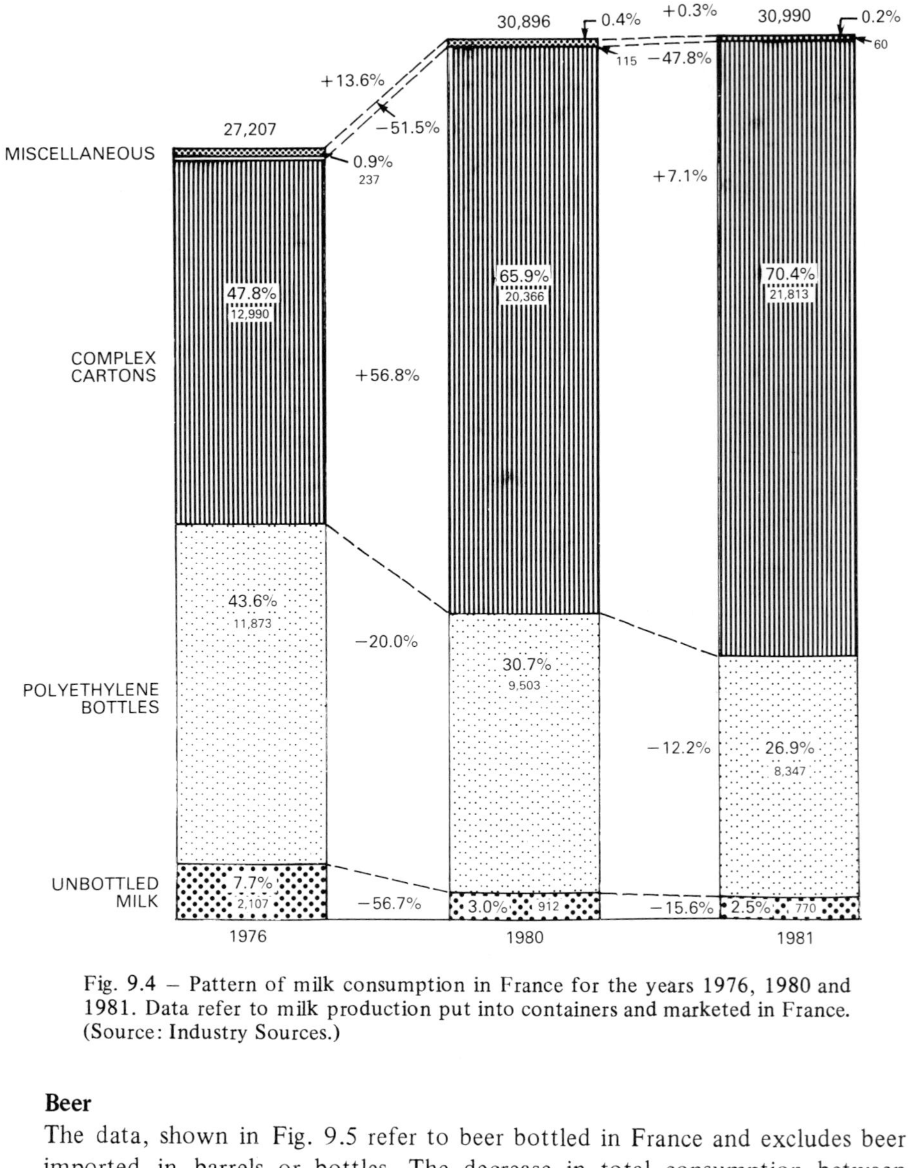

Fig. 9.4 – Pattern of milk consumption in France for the years 1976, 1980 and 1981. Data refer to milk production put into containers and marketed in France. (Source: Industry Sources.)

Beer

The data, shown in Fig. 9.5 refer to beer bottled in France and excludes beer imported in barrels or bottles. The decrease in total consumption between 1976 and 1980/81 can be explained by the exceptionally warm summer of 1976. In fact, beer sales are gradually increasing (21.5 million hectolitres in 1981 as compared with 20.6 million hectolitres in 1980).

Sales of beer in 30- and 50-litre barrels (draught beer) are expanding (23.4% of the total sales in 1981 as compared with 18.5% in 1976). Sales in non-returnable glass bottles increased substantially between 1976 and 1980, but this now seems to be slowing down. Returnable glass bottles are less used than in previous years (42.3% of total sales as compared with 58% in 1976). The use of cans is rapidly growing (140 thousand hectolitres in 1981 as compared to 81 thousand hectolitres in 1980) but sales in cans are still insignificant (0.6% of total sales).

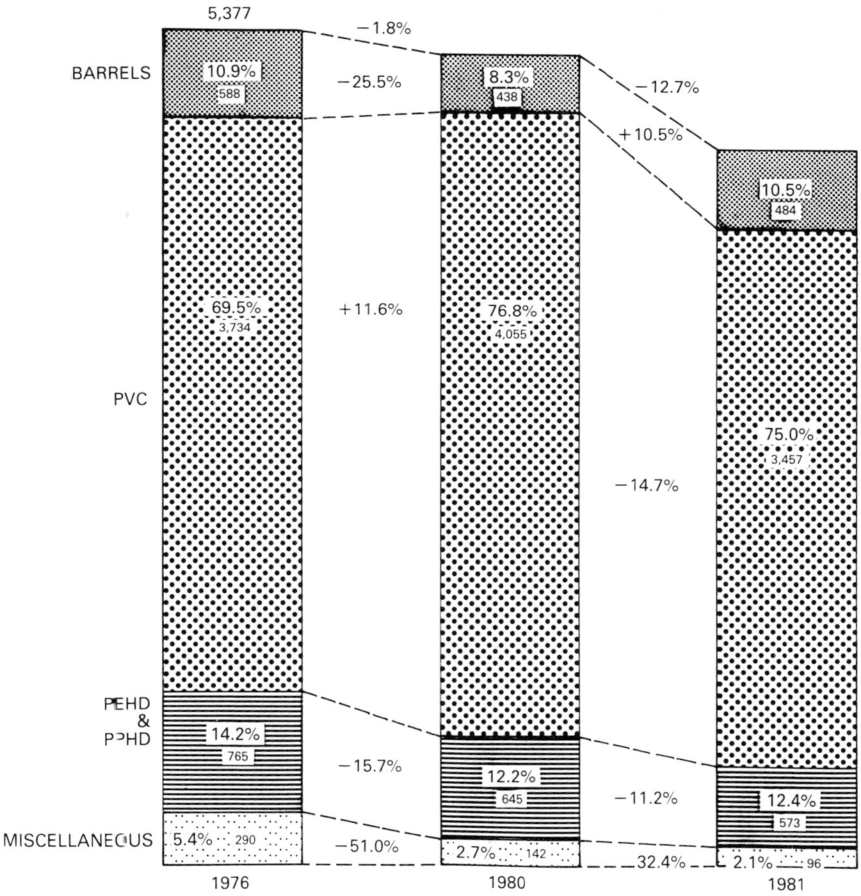

Fig. 9.5 – Pattern of beer consumption in France for the years 1976, 1980 and 1981. (Source: Union Generale de la Brasserie Francaise (UGBF).

Panache (Shandy)

Until 1979, 'Panache' was only consumed in cafés where beer and lemonade on tap were poured together. Sales of bottled 'Panache' are rapidly expanding, but they remain at a low level of consumption (about one litre per person per year) and are almost exclusively in non-returnable glass bottles (96% of total sales) as shown in Fig. 9.6.

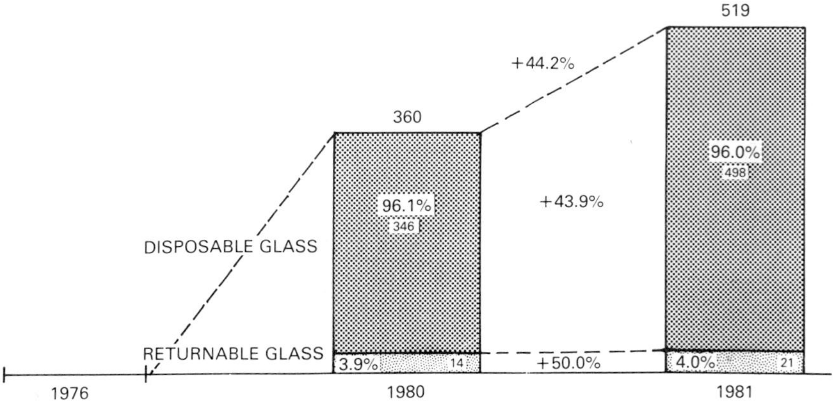

Fig. 9.6 – Pattern of panache (shandy) consumption in France in the years 1976, 1980 and 1981. (Source: Union Generale de la Brasserie Francaise.)

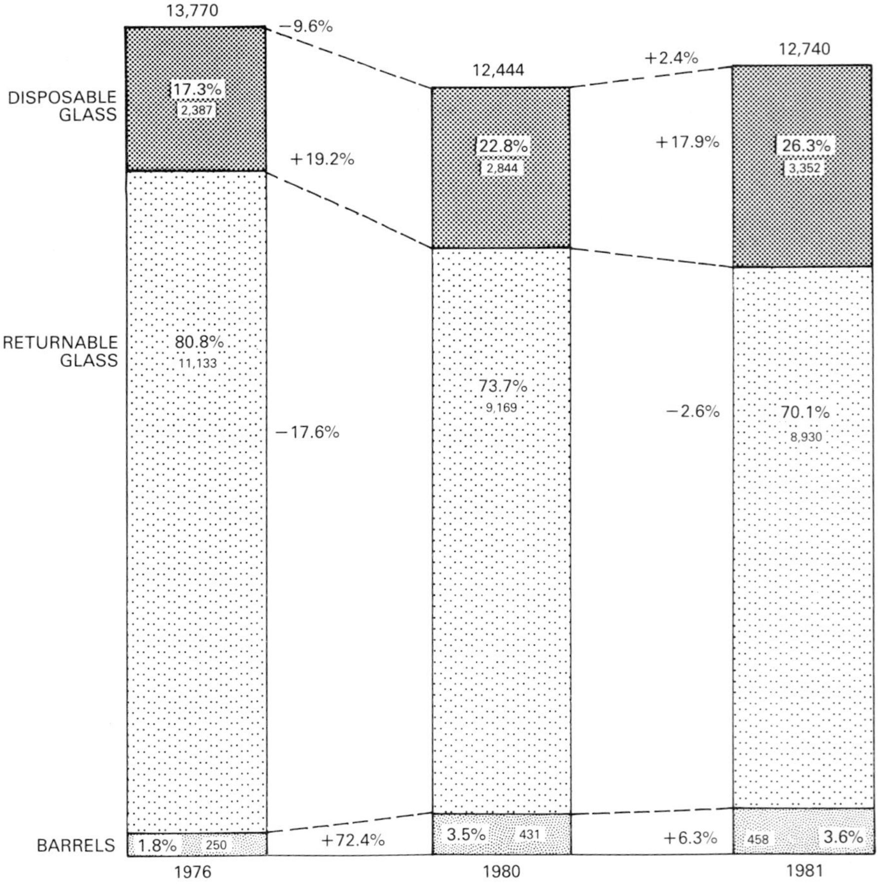

Fig. 9.7 – Carbonated soft drinks consumption in France in 1976, 1980 and 1981. (Data based on industry sources.)

Carbonated soft drinks

The consumption of carbonated soft drinks is shown in Fig. 9.7. Sales of soft drinks are slowly increasing (12.7 million hectolitres in 1981). Sales of soft drinks in 30- and 50-litre barrels (on tap) to cafés and cooperatives are expanding. After a significant decrease between 1976 and 1980, sales in returnable glass bottles are now stabilising as are sales in non-returnable glass bottles.

Edible oil

Over the last two years, the consumption of edible oil has decreased significantly as shown in Fig. 9.8 from 5.3 million hectolitres in 1980 to 4.6 million hectolitres in 1981, i.e. a drop of 12.7%. Annual sales figures only relate to wholesale sales and do not take into account the variation of inventories in the distribution network. All oil is sold in non-returnable containers, including metal barrels. After an increase in the use of PVC containers between 1976 and 1980, this kind of container (1- to 3-litres) has seen a slight decrease in usage. The one-litre PVC container remains the most commonly used (68% of total sales).

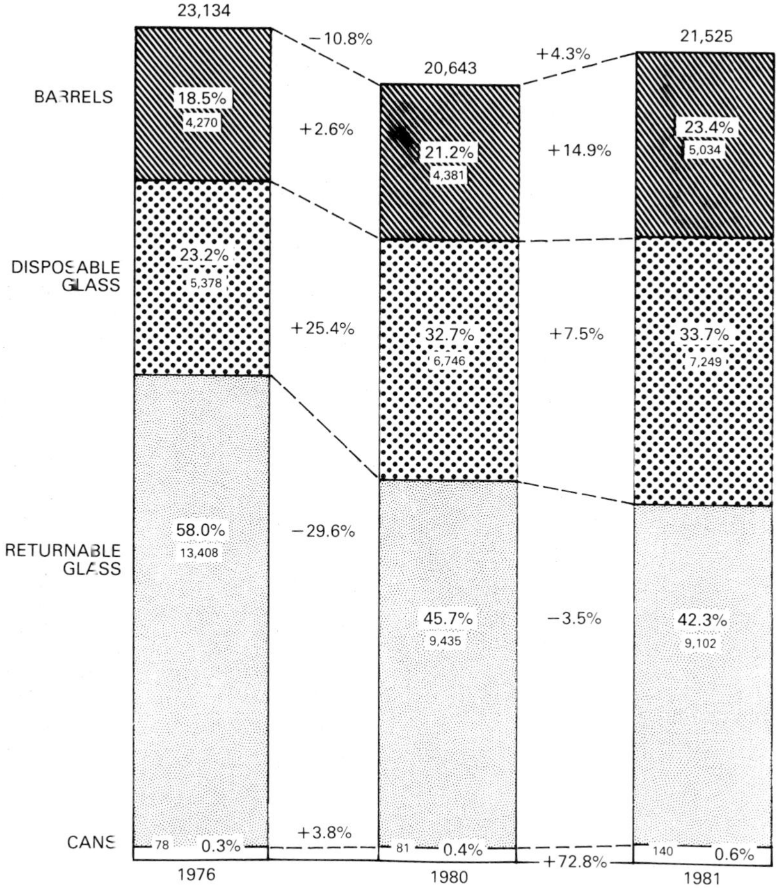

Fig. 9.8 – Consumption of edible oil in France in 1976, 1980 and 1981. (Source: Syndicat des Petites et Moyennes Enterprises d'Huileries.)

The use of very large containers (3- to 5-litre containers in PVC or polyethylene) is increasing.

Aperitives and Spirits

The sales of aperitives and spirits are shown in Fig. 9.9 and relate to wholesale sales; they do not take account of any changes in stock in the distribution network. In 1981, the total consumption of alcoholic beverages amounted to 3.7 million hectolitres, or some 7 litres per person per year. About 13% of total consumption is represented by bottle imported products such as whisky, gin etc. and this figure is increasing. Recent increases in the use of 70-centilitre bottles rather than 1-litre bottles can be explained by the price increases of these beverages.

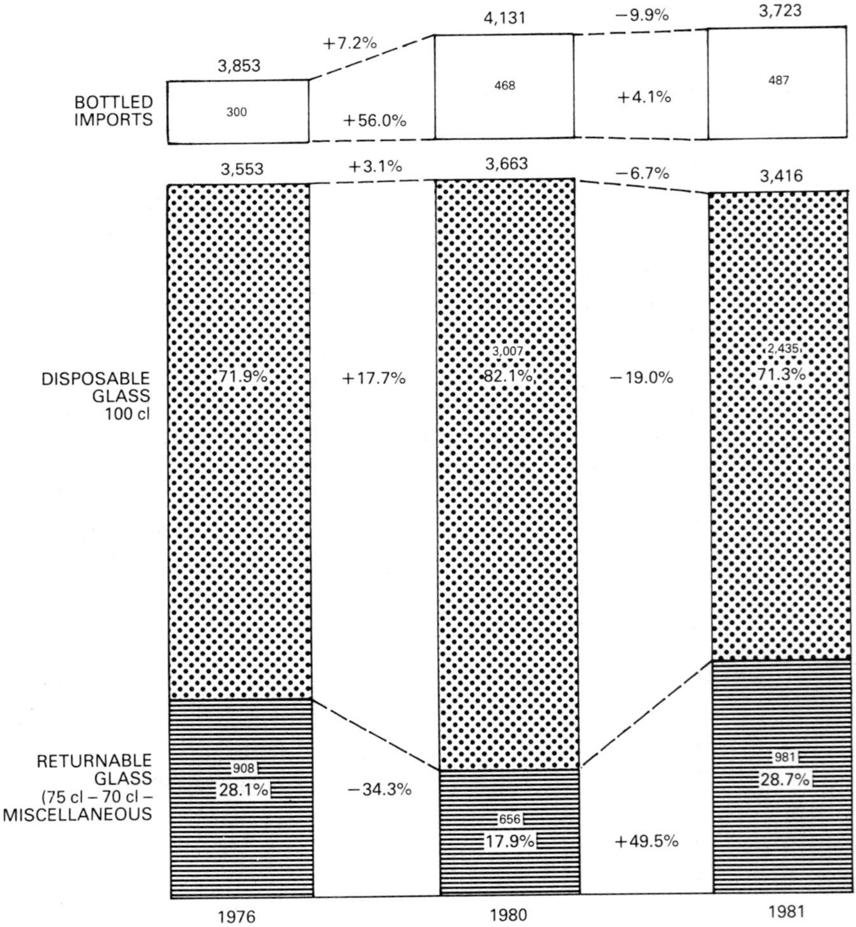

Fig. 9.9 – Consumption of spirits and aperitifes in France in 1976, 1980 and 1981. (Data based on industry sources.)

Fruit juices

Consumption of fruit juice is rapidly expanding (1.7 million hectolitres in 1981 as compared with 1.2 million hectolitres the year before, an increase of 35%). The professional bodies explain that this spectacular increase is partly attributable to the introduction of new containers ('briques' – board and polyethylene) while sales in other traditional containers remain stagnant. It should be noted that there are practically no sales of frozen fruit juices in France.

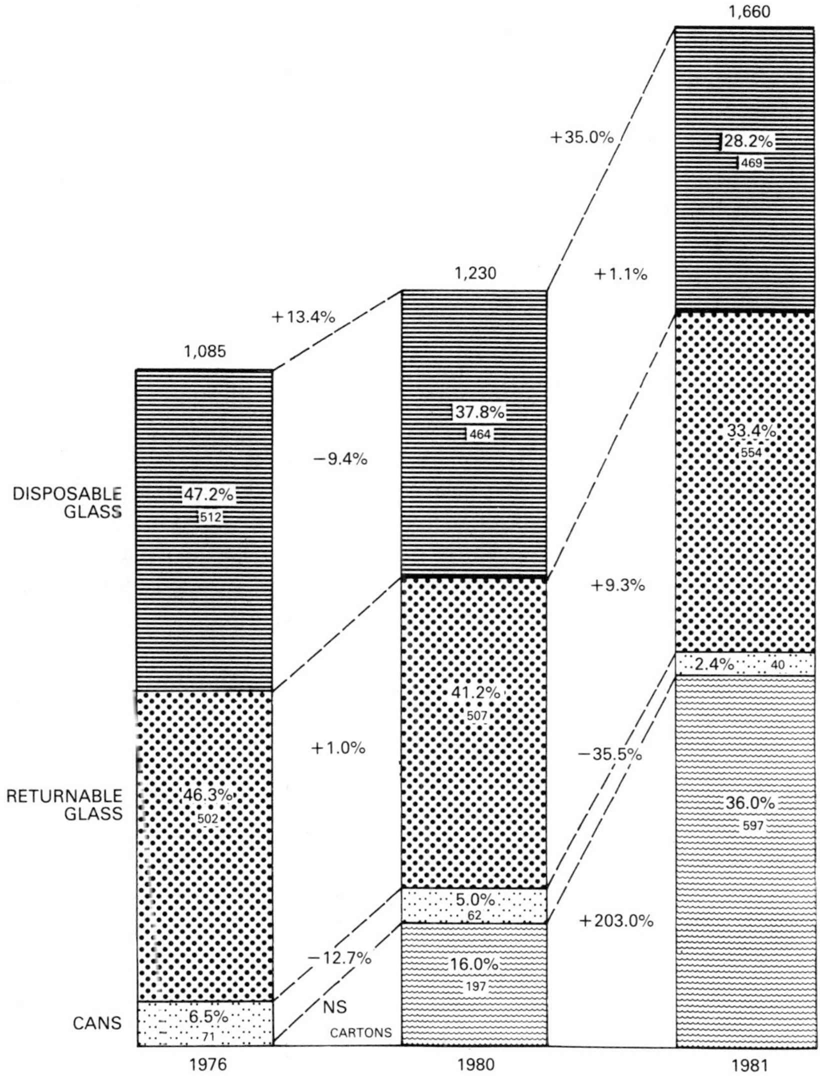

Fig. 9.10 – Consumption of fruit juices and cordials in France in 1976, 1980 and 1981. (Source: Union des Jus de Fruits.)

Three points should be noted:

(a) the spectacular increase in the sales of fruit juice in 'briques' (36% of total sales in 1981);
(b) the stable sales position of fruit juice in returnable or non-returnable glass bottles;
(c) the decrease in sales of fruit juices in cans.

About two-thirds of French, domestically-bottled, beverages for home consumption are sold in non-returnable containers, as indicated in Table 9.2. Amongst returnable containers, the proportion of returnable glass bottles is close to 90%. For non-returnable containers, three materials have a significant share of the market: glass (31%), plastic (PVC and polyethylene 34%) and board (33%).

The sales of beverages to cafés, hotels restaurants, and cooperatives amount to 23% of total sales. Beverages sold through food stores represent 77% of total sales; of this amount, 30% are sold through traditional shops and local stores and 70% through 'modern' distribution channels (hypermarkets, supermarkets and superettes). The detailed breakdown is shown in Table 9.3.

THE OBJECTIVES TO BE ACHIEVED BY 1984

As already noted, 1976 was taken as a reference year for energy consumption and waste production. In 1976, domestic sales of beverages bottled in France amounted to 149,231 thousand hectolitres. The energy required for bottling, distributing, disposing or recycling the containers amounted to 11,582 million therms or 1,203 thousand tonnes of oil equivalent. The existing sales mix, including returnable and non-returnable containers, produced 1,903 thousand tonnes of waste. Accordingly, the reference indexes were:

Energy: 8.063 kg oil equivalent per hectolitre bottled and sold in France (or kg . oe/hl)
Waste: 12.75 kg waste per kectolitre of beverages (kg/hl)

As defined in the text of the agreement signed on 17 December 1979 between the government and industry, the objectives were a reduction of 12% in energy consumption and 40% in waste production, to be attained five years hence, i.e. in 1984. In other words, the objectives were to achieve an energy consumption of $8.063 \times 0.88 = 7.095$ kg . oe/hl and a production of waste of $12.75 \times 0.06 = 7.65$ kg/hl.

RESULTS ACHIEVED IN THE FIRST TWO YEARS

The model to be used in monitoring the progress, the definition of data to be collected, drafting of forms to be used, implementation within a sampling of

Table 9.2

Breakdown of beverage sales in France by type of container in 1981. All data are in thousands of hectolitres

Beverage	Returnable containers			Non-returnable containers				Totals
	Barrels	Glass bottles	Misc.	Glass bottles	Plastic bottles	'Bricks'	Misc.	
Wine	–	27300	363	12125	1810	–	1287	42885
Mineral & spring water	–	3383	–	1097	26100	–	–	30580
Beer	5034	9102	–	7249	–	–	140	21525
Panache (shandy)	–	21	–	498	–	–	–	519
Milk	770	–	–	60	–	30160	–	30990
Soft drinks	458	8930	–	3144	–	–	208	12740
Edible oil	–	–	–	–	4030	–	580	4610
Alcoholic beverages	–	–	–	3416	–	–	–	3416
Fruit juices	–	554	–	469	–	597	40	1660
Totals	6262	49290	363	28058	39194	30757	2255	148925
		55915			93010			

Table 9.3

Breakdown of beverage sales in France in 1981 by type of retail outlet. All data are in thousands of hectolitre

Beverage	Cafés, hotels restaurants, cooperatives	Food stores				Total
		Hypermarkets	Supermarkets	Superettes	Traditional & local stores	
Wine	15130	6090	6060	5520	9905	42885
Mineral & spring water	2543	9399	9029	3996	5613	30580
Beer	9223	2919	3843	2112	3428	21525
Panache (shandy)	19	210	184	66	40	519
Milk	3161	5705	7013	4341	10770	30990
Soft drinks	2420	2225	2936	1826	3333	12740
Edible oil	875	1251	1446	403	635	4610
Alcoholic beverages	954	1027	827	342	266	3416
Fruit juices	205	614	374	227	240	1660
Totals	34530	29440	31712	18833	34230	148925
		114215				

French companies, etc. were set up in 1980, and an operational test was conducted despite the fact that the results had not been taken into account in the text of the signed agreement. Two further real runs were effected in 1980 and 1981. The results registered on these three runs are shown in Table 9.4 for energy consumption and Table 9.5 for waste production.

Table 9.4

Estimate of energy consumption using the model set up to monitor the French agreement

	Years			
	1976 reference	1979 test	1980	1981
Volume of sales in 1000 hectolitre	149231	146842	147016	148043
Consumption of energy in 10^6 therms	11482	11644	11272	10950
Ratio of energy kg o.e./hl	8063	8238	7966	7684

Table 9.5

Estimate of waste production using the model set up to monitor the French agreement

	Year			
	1976 reference	1979 test	1980	1981
Volume of sales 10^3 hectolitre	149231	146842	147016	148043
Waste production 10^3 tonne	1903	1984	1941	1856
Ratio (kg/hl)	12.75	13.51	13.20	12.54

After 1976, energy consumption per hectolitre increased until 1977/78 and then slowly decreased. In 1979, the energy index was 8.238 kg . oe/hl compared with the value of 8.063 kg . oe/hl in 1976. At that point, the index showed a definite trend downwards, mainly due to the efforts of container manufacturers in recycling non-returnable containers in greater quantities.

In 1981 the energy index stood at 7.684 kg . oe/hl compared with the 1976 value of 8.063 kg . oe/hl; i.e. a decrease of 4.7%. Waste production in the beverage sector followed the same pattern as energy consumption, but with a time lag of one to two years. In 1979 the waste index was higher than the reference figure at 13.5 kg/hl compared with the 1976 value fo 12.75 kg/hl. The index then started to decrease slowly reaching 12.54 kg/hl in 1981.

To attain their objectives within the next three years, the relevant sectors of industry must reduce their energy consumption by 7.67% over the next three years, which means an annual decrease of 2.6%. In addition they must decrease waste production by some 61% over the next three years, which means an annual decrease of 15.6%. Figures 9.11 and 9.12 show the registered trends and the efforts needed to be on target by 1984.

DISCUSSION OF THE CHANGES

Energy consumption for bottling different beverages varies considerably from 4.578 kg . oe/hl for milk to 19.515 kg . oe/hl for alcoholic beverages based on

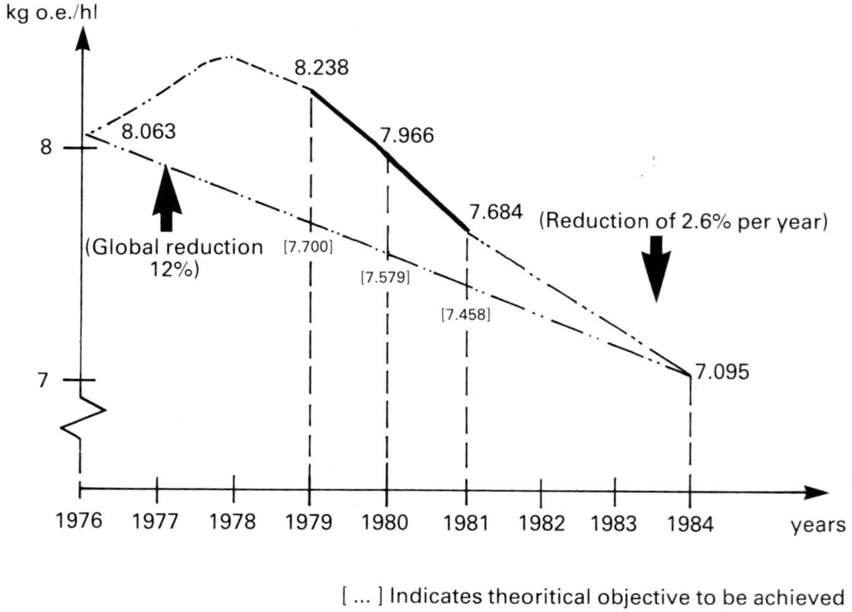

Fig. 9.11 – Energy consumption index in kg oil equivalent per hectolitre (kg oe/hl) as a function of time. The solid curve shows actual achievements; the broken lines show the progress that must be made to reach the target 1984 values specified in the agreement.

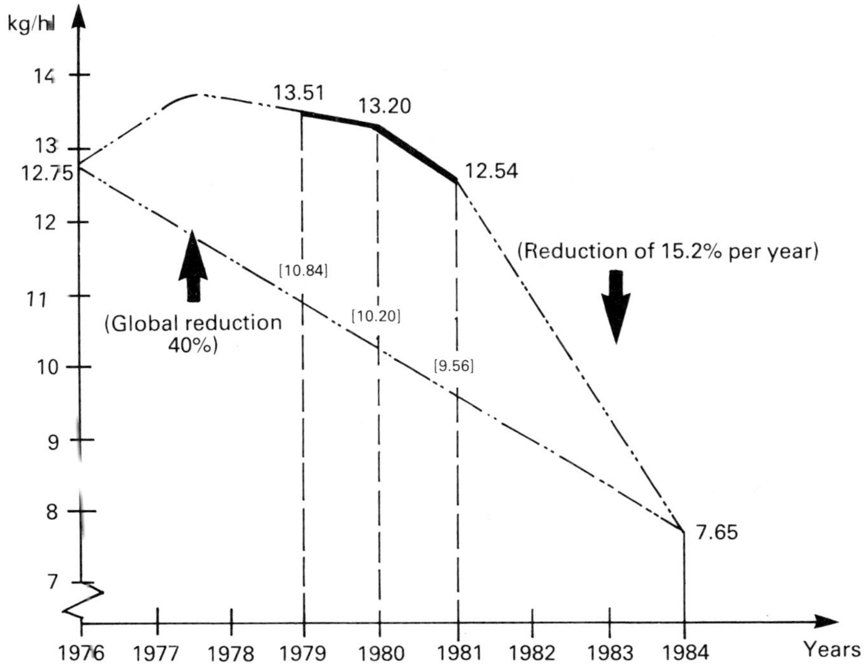

Fig. 9.12 – Waste production index in kg per hectolitre (kg/hl) as a function of time. The solid curve shows actual progress; the broken lines show the progress that must be made to reach the target 1984 values specified in the agreement.

1981 results, with an average of 7.684 kg . oe/hl. The decrease of 4.7% since 1976 results from two opposing elements. On the one hand, market demand for more products in disposable containers has increased, thus consuming more energy and producing more waste, whereas on the other hand, efforts by professional sectors to decrease the energy consumption, and, to a lesser extent, waste production have met with some success.

It should be noted that as the signed agreement relates to all beverages, no precise details concerning energy consumption or waste production for a specific product should appear in external publications. Only qualitative details can be mentioned. Therefore, compared to their 1976 reference figures – on a product by product basis – beer and soft drinks still appear as large energy consumers, while a decrease in energy consumption has been noted for other products. This deterioration in the energy index for beer and soft drinks results from a decrease in sales in returnable glass bottles (usually with a very high trippage), an increase in sales in non-returnable containers and the introduction onto the market of the new 'panache' (shandy), which is practically all sold in small, non-returnable glass bottles. The trend however is now stabilising as market demand reveals a return to the old position regarding non-returnable small bottles, after a period of increasing sales.

For all other beverages, the energy index decreased but with variations from product to product. For wine, sales in PVC bottles decreased while sales in larger containers (e.g. polyethylene polypins) increased. There was also a stabilisation of sales in 75 cl bottles between non-returnable and returnable glass bottles. For mineral and spring waters, the old, 90 cl bottles were replaced by lighter, 1-litre bottles. Milk sales changed from polyethylene bottles to cardboard and polyethylene bricks, which are less energy intensive. The main change in edible oil sales is the move to larger containers and to non-returnable barrels. For spirits and aperitives, there has been a marked decrease in weight of non-returnable glass bottles as well as reduction in the use of cardboard outer packaging. For fruit juices, the main change is to sales in 'briques'. The overall effect of these changes in market demand is an increase of 10.8% in energy consumption per bottled hectolitre, compared to the 1976 situation. These effects are summarised graphically in Fig. 9.13.

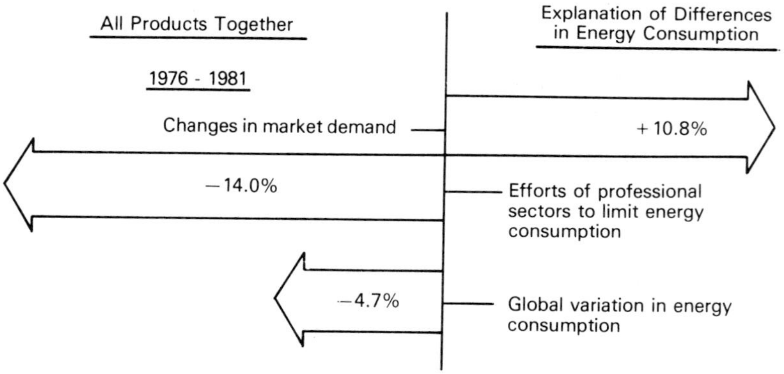

Fig. 9.13 – The effects of changes of market demand and efforts of industrial sectors on overall energy changes 1976–1981.

The efforts of the industrial sectors to limit energy consumption have been concentrated in three main areas:

(a) a general decrease in the weight of containers,
(b) increased use of wastes, mainly by recycling cullet, and
(c) improved production processes in the glass and plastics industries.

In general all kinds of containers were lightweighted between 1976 and 1979. The weight of barrels decreased by about 1%. The weight of PVC bottles used for wine and mineral waters decreased, whereas the weight of containers for edible oil increased. Overall, a decrease of 2.2% has been registered for PVC bottles. The weight of polyethylene containers for milk and edible oil decreased by 3.3% and 11.4% respectively over this period, and a slight decrease has been noted in cardboard containers (5.4%). The most significant decreases however have been achieved in the weight of glass containers; 4.4% for returnable bottles and 15.8% for non-returnable bottles. The overall result of these lightweighting measures has been a 5.2% saving in energy compared with the 1976 situation.

The increased use of waste included re-use of bottles, recycling of cullet, regeneration of PVC and incineration of waste (including cardboard, PVC and polyethylene, with heat recuperation. Some 116 million bottles are now re-used annually, of which 110 million are wine bottles and 6 million are beer bottles. In 1979, only 92 million bottles were re-used.

A spectacular increase has been noted in cullet recycling. In 1981 some 416,115 tonne of cullet was recycled compared with 336, 000 tonnes in 1980 and 277,000 tonnes in 1979. At the end of 1981, cullet was collected in some 8,544 districts, compared with 6,266 districts in 1980. 26.4 million inhabitants were involved in returning glass bottles in 1981, as compared with 21.4 million in 1980, an increase of 24%, resulting in an additional 270, 000 tonnes of glass. In 1981, 10.2 kg of glass per person year were collected as compared with 8.6 kg in 1980. Glass manufacturers are continuing their efforts.

Because of the difficulty in collecting empty PVC bottles at a realistic cost, and in producing high quality regenerated PVC material and marketing it, only 1,130 tonnes of PVC were regenerated in 1981, as compared with 1,000 tonnes in 1980.

The incineration of waste with heat recuperation covers about 25% of the waste material collected.

The combined effect of all these factors resulted in a saving of 3.7% in energy consumption in 1981 compared with that in 1976.

Energy savings were noted in technological processes in the glass and plastic industries. Compared with 1976, the glass industry saved some 2.5% and the plastics industry saved approximately 1.7%.

It is clear from this discussion that several factors are responsible for the changes in energy consumption which have been identified and evaluated. Other parameters, which are quantifiable to a greater or lesser extent, could affect the overall energy index in one direction or another. Amongst these are:

(a) changes in the use of outer packaging (e.g. replacing cardboard by polyethylene film),
(b) changes in physical distribution (e.g. greater or lesser distances covered in delivery rounds and the importance of direct deliveries),
(c) changes in bottling speeds, and
(d) changes in trippage, or in distribution patterns.

To sum up then, the change in energy consumption used in the bottling and distribution of beverages in France between 1976 and 1981 are:

Increase due to changes in market demand	+10.8%
Decreases due to	
lightening of containers	–5.2%
use of waste	–3.7%
improvements in manufacturing technology	–2.5%
miscellaneous other reasons	–1.7%
Overall variation	–4.7%

These figures are illustrated in Fig. 9.14.

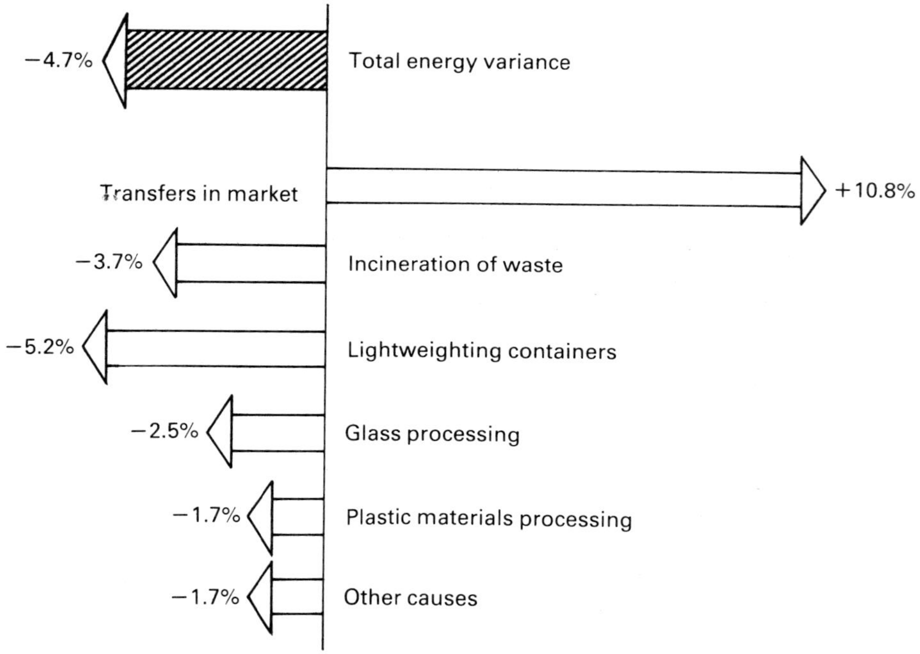

Fig. 9.14 – Changes in energy consumption for all beverage container products 1976–1981 showing the origin of the different components.

ACKNOWLEDGEMENTS

This presentation was made possible by courtesy of the French sponsors of this study: The Agence Nationale pour la Récuperation et l'Elimination des Déchets (ANRED) – (on the government side represented by Mr. Michel Affholder, Directeur); and the 'Groupement d'Etudes et de Normalisation des Fabricants

(GENFA)' (for industry represented by Mr. Jacques de Pastors, Managing Director and Mr. Henri Giradel, Chairman of the Association 'Progress et Environment'). All data and figures included in this section have been extracted from various reports issued by Peat, Marwick, Mitchell & Co., Consultants (France). They are the joint exclusive property of ANRED and GENFA.

The author

Jean-Claude Cohadon graduated in Agricultural Economics and Alimentary Technology from the Institut National Agronomique, Paris. He then pursued studies at Cornell University, Ithaca, New York and graduated with an M.Sc. in Economics. In 1963 he joined the Compagnie Francaise d'Organisation (COFROR), a French consulting firm specialising in the food industry and marketing. In 1973, at the merging of COFROR with Peat, Marwick, Mitchell & Co. Consultants, he joined the PMM Organisation. Mr. Cohadon has been a partner of Peat Marwick International since 1979 and Partner-in-Charge of the French Management Consulting Practice of Peat, Marwick, Mitchell & Co. Since 1981.

CHAPTER 10

Recycling of aluminium cans – from voluntary system to public intervention

MIKAEL BACKMAN
TEM, Lunds Universitet, Sweden

BACKGROUND

Beverage containers have been the subject of public debate in Sweden ever since the mid-1960s. At first, most criticism was levelled at the increasing use of non-returnable glass bottles, which were thought to have considerable drawbacks, such as the risk of personal injury. These criticisms are probably one reason why non-returnable glass containers now hold a very limited market share.

However, as the non-returnable glass bottle disappeared from the market, it was replaced by another non-returnable container – the can. As a result, the proportion of non-returnable containers has held fairly steady at around 35% in the past 15 years. This can be seen from Table 10.1, which also shows that Swedish consumption of beer and soft drinks fell off in the second half of the 1970s.

A major change that has occurred in the can sector is displacement of the three-piece steel can with an aluminium end by a two-piece, seamless, all-aluminium can. The present paper deals with this all-aluminium container.

HISTORY OF THE ALUMINIUM CAN

Aluminium cans have been on the Swedish market since 1977. As more and more breweries imported this cheaper package instead of buying the Swedish-made steel can, the aluminium can increased its share of the can market from 11% in 1978 to 28% in 1980. However, the debate on cans did not get under way until 1980, when it became clear that the Swedish can manufacturer, PLM, intended to compete with foreign suppliers. With that aim in mind PLM built a new factory in Malmö, meanwhile phasing out its steel can line. Since coming onstream in 1981, the new factory supplies the Swedish brewing industry with about 500 million cans a year, which means that almost all cans being sold at present are made of aluminium.

Hence, in the space of five years, the beverage can market has switched totally from steel to aluminium cans. Even without public intervention, this development would most likely have occurred even if aluminium cans had not been manufactured domestically. But as already mentioned, at the same time as it was clear to consumers and politicians that a total switchover to aluminium was about to occur, a fierce discussion of the new container flared up. Since aluminium is an expensive raw material and an attractive residual product, much of the debate obviously centred on suitable ways of recovering the cans once they were emptied of their contents. Ultimately, this resulted in a parliamentary decision on recycling via a system of mandatory deposits (of which more below). First, however, it is of interest to comment on the processes and arguments which finally led to the intervention of central government.

SYSTEM OF RETURNS – FORMULATION OF GOALS

Figure 10.1 illustrates the interactions in a system of returns which could underlie studies of various alternatives concerning both recycling activities and products on which such activities focus. This approach can accordingly subsume returnable as well as non-returnable containers.

The retrieval rate (i.e. the extent to which the products in question are returned) is a central factor in every system of returns and the factors which affect retrieval rate are:

(a) the collection system (whose design depends on available technology, as well as retrieval station and collection costs),
(b) information (whose volume and design are governed by economic considerations and by the target audience) and
(c) repurchase value of the container, which is basically dictated by the price of scrap, and this, in turn, is directly related to the price of natural raw materials and to collection costs. Political decisions or other influences can be used to raise the repurchase value above this basic value. The repurchase value of returnable containers is determined by the deposit, which in turn is limited by the costs of manufacturing inclusive of container levy, if any.

The central factor, retrieval rate, can be assessed with reference to the result factors (littering, resource use, waste generation and cost-benefits) by relation to a specified zero state for these factors. It is evident that the result can give rise to changes in certain fundamental and contributory factors.

The 'attitudes and values' of the various interests involved have a blanket effect on the design of the system, on factors at different levels and on the relations between these, and also on the overall evaluation of the system. The term 'interests' is understood to mean all parties who come into contact with

Table 10.1
Consumption of beer and soft drinks in Sweden broken down by container type

	Year					
	1965	1967	1972	1975	1977	1979
Total consumption of beer and soft drinks 10^6 litre	538	751	748	774	717	699
% sold in returnable containers	84	68	63	68	64	64
% sold in non-returnable containers	16	32	37	32	36	36
% non-returnable containers that are cans	6	50	54	81	87	92
% other non-returnable containers	94	50	46	19	13	8

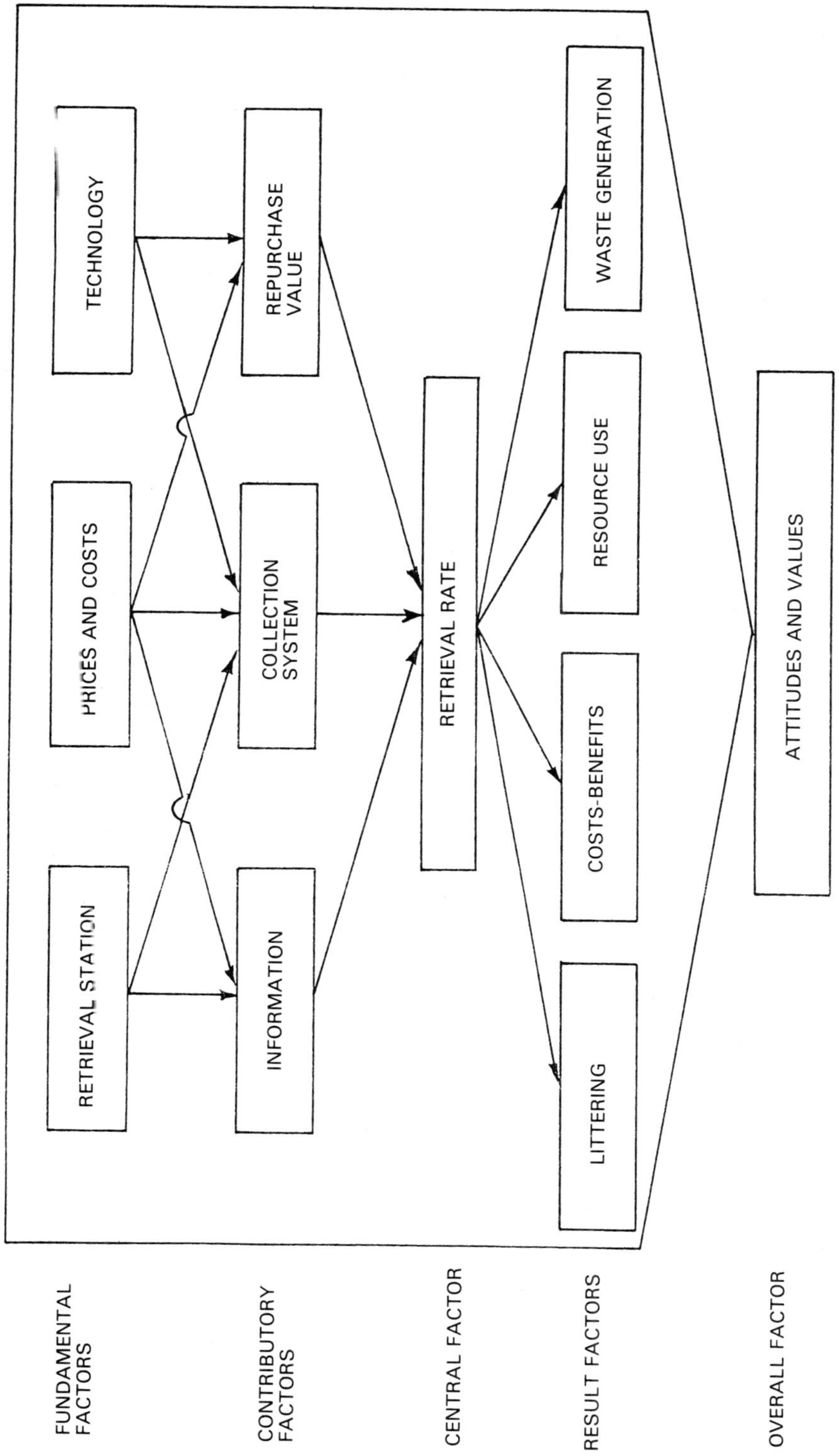

Fig. 10.1 – System of returns – system factors and interrelations.

the system in one way or another, among them manufacturers, consumers, government agency officials, politicians, retailers, personnel in charge of handling activities, people responsible for refuse collection and disposal etc.

A higher degree of recycling is often portrayed as a special goal towards which society ought to aspire. However, recycling is not a goal in itself but a means of attaining certain other goals. This means that the desired goals must first be formulated, then those goals that are achievable should be identified. In order to reach various goals, recycling will sometimes be found among the possible means, but this does not imply that recycling should invariably be chosen. Instead, the different approaches to the problem should be weighed against one another and that which permits attainment of the goal at minimum macroeconomic cost should be chosen.

It follows therefore that the central factor in a system of returns, the retrieval rate, cannot *per se* be regarded as any measure of the fulfilment of the goal of the system. Instead the goals must be clearly and conclusively formulated *in advance* in such a form that they can be converted into a definite retrieval rate. Usually however, public goals are seldom formulated in absolute terms; if anything, they indicate a direction of will. So, the question of whether or not different systems of returning beverage containers are appropriate means for attaining desired results cannot be resolved by a simple appraisal, because the goals themselves are only vaguely defined. Instead, the goals must be circumscribed as far as possible and the various systems will have to be appraised in relation to these goals and also in relation to other means using macroeconomic considerations.

One way of specifying public goals for the beverage container sector may be to begin with those arguments which are put forward in public debate and to use these as premises. In general, the commonly accepted goals would be:

(a) to reduce the pollution of water and air from the beverage container sector,
(b) to reduce the consumption of energy and raw materials in the beverage container sector,
(c) to reduce consumer costs, inputs of labour and other sacrifices,
(d) to reduce the waste generated by the beverage container sector, and
(e) to reduce littering by beverage containers.

Clearly these goals are common not only to the packaging sector but to many other industrial sectors, indicating that a policy which seeks to direct the course of events towards the goals should preferably be wide-ranging rather than attack products.

Returning to the goals, however, it was observed that in relation to aluminium cans in 1979–1981, there was wide disagreement as to the retrieval rate society would demand of a system of returns. Several exact percentages were given without these being related to any goal. In those cases where different goals were discussed, the disagreement extended to what goal or goals the retrieval rate should be related to. Finally, the value of retrieval rate also varied because of disagreement over whether the aluminium can should be compared with the conventional tin can or with returnable glass bottles. In the summer of 1980, an official committee was appointed to inquire into these matters. At the same time, PLM and the state-owned brewing firm, Pripps, ran an experimental pilot scheme on collecting used cans.

This trial, called 'Project Recan', was implemented in Varberg on the west coast of Sweden. It closely resembled the collection trials mounted by several other countries, except that the collection of cans was linked with an already established system for collecting newspapers and periodicals. Before considering the results of the official committee's investigation, we shall briefly describe this system of returns.

'PROJECT RECAN'

The system of returns that PLM and Pripps tested in Varberg was based on voluntary consumer participation and on the tare value of the scrap metal. It consisted of three main components:

(a) *Direct collection from households*
Such collections were carried out monthly in connection with the collection of newspapers and periodicals.

(b) *Emptying receptacles intended for empty cans*
These receptacles were placed at strategic sites in the municipality, for example, in parks, in kiosks and fast-food outlets, on town squares, market-places, etc.

(c) *Recurrent clean-up campaigns*

Table 10.2 gives the number of cans delivered into the area and the number of collected cans during the period of time allotted for the trial.

Table 10.2

Number of cans shipped-in and collected in Varberg during February–October 1980. Note that collections started in mid-March

Month	Number of cans shipped in	Number of cans collected	Number of cans from clean-up plus purchases
February	126100	–	–
March	156800	10100	–
April	166200	34700	–
May	196700	54800	–
June	312400	119300	–
July	372600	103300	–
August	244500	62000	–
September	163100	63500	–
October	146500	70500	21000

The number of cans shipped in was calculated from the sales statistics of the largest breweries. The number of collected cans was determined by both manaul counting and control weighings. A clean-up campaign undertaken in October brought in 8,000 cans. In the same month, cans were repurchased from the general public for 5 öre each, i.e. the scrap value. In total, these two campaigns brought in 21,000 cans during the month of October. The time delay between retail delivery, consumption and collection cannot be determined with any certainty. However, an average of about two weeks is estimated for transport time and shelf time in stores. When the storage time in households is added, it is reasonable to assume a time delay of one month. If the extra collection of cans in October is spread out evenly over the whole trial period, the result is a retrieval rate as a function of time as shown in Fig. 10.2. It will be seen from this diagram that the retrieval rate drops sharply in the summer months. Collections again improve with the onset of autumn and reach the early summer level in October.

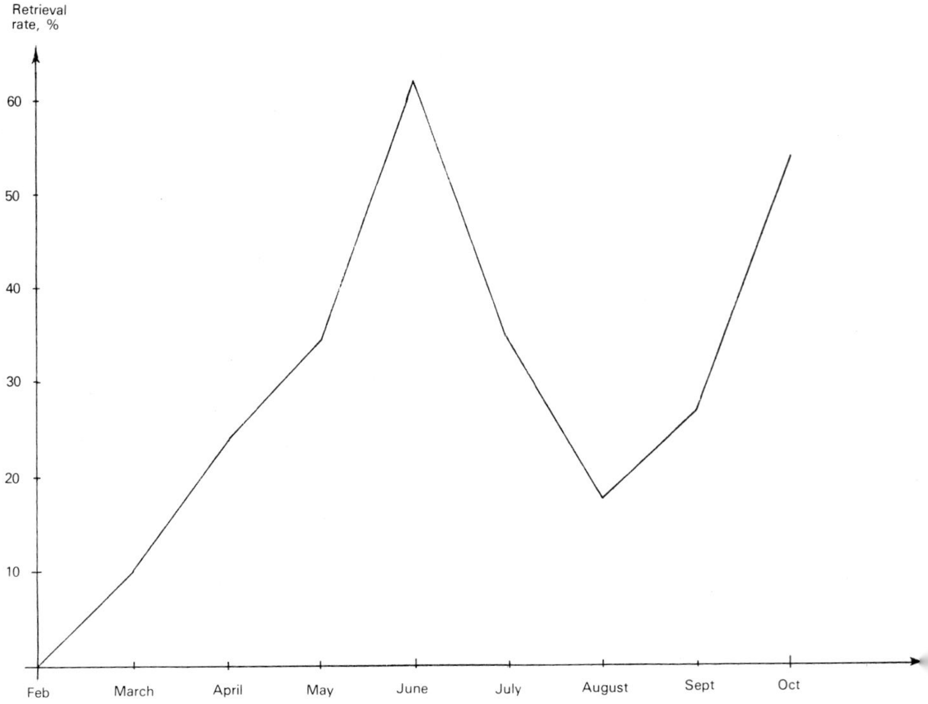

Fig. 10.2 – Retrieval rate as a function of time for the recovery of cans in the Varberg experiment.

In connection with the recycling trial, four market surveys were carried out. Among their aims were to chart knowledge of and attitude to the trial. Table 10.3 shows the responses given to the directly asked question: 'Do you know anything about "Operation Recan"?'.

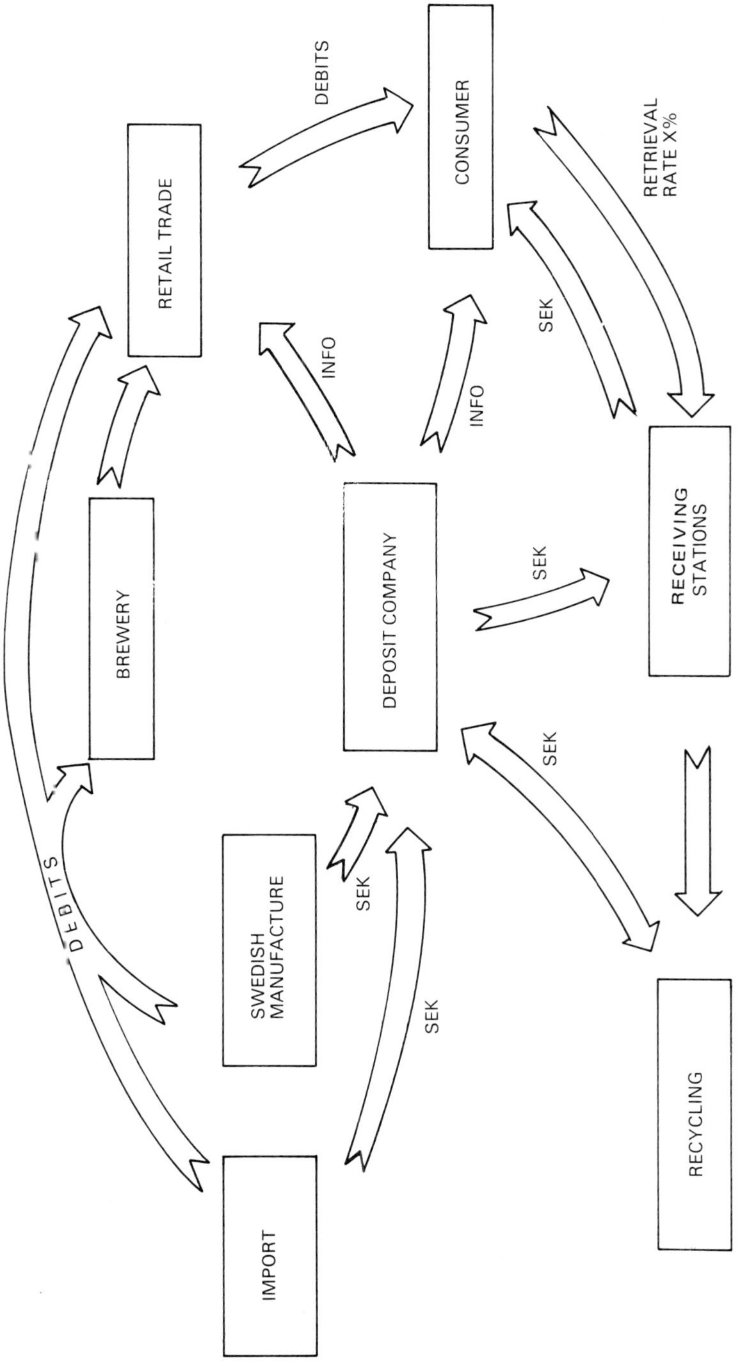

Fig. 10.3 – The deposit system proposed by TEM and eventually adopted by the Swedish government for handling aluminium cans.

Table 10.3

Knowledge of 'Operation Recan' among the year-round population and among the summer population. All values are percentages

	Year-round (April to April)	Summer population		Year round population	
		Can consumers	Not can consumers	Can consumers	Not can consumers
Do *not* know of the scheme	10	44	40	4	8
Know of the scheme	90	56	60	96	92

The table shows that knowledge of 'Operation Recan' among the year-round population maintained a high level throughout and also improved somewhat as the trial progressed. However, such knowledge was much lower among the summer population, which is unfortunate because there is a substantial influx of tourists to Varberg in the summer months. It is estimated that the year-round, settled population (of 44,000) more than doubles at regular intervals, while *on average* the summer population causes an increase of about 20,000. Another reflection of the sharply increasing population in summer is that the number of beverage cans delivered to retailers in the area more than doubled during June and July. In contrast, the number of collected cans did not increase to the same extent, which among other things resulted in a low retrieval rate for August (about 20%).

In April, the permanent population were asked whether they intended to take part in the project. The same question was put to the summer population in June. Thus, by the time of the October survey, it was possible to check how many people really followed through on their intentions. The results are set out in Table 10.4. Since the expressed *intention* to participate is not combined with

Table 10.4

Results of a survey of can consumers who intended to participate and who did participate in the can collection drive. All values are percentages

	Year-round population (April)	Summer population	Year-round population (October)
Intend to participate and did participate	86	81	60
Do *not* intend to participate and did *not* participate	14	19	36
No answer	–	–	4

any responsibility for practical action, it is inevitable that some apathy should be recorded after the trial had been going for some time and the consumers had acquired practical experience of their assistance. Only a few present of the respondents said they had collected fewer cans due to the massive criticism levelled at the trial via the press, radio, TV, leaflets and open-air theatre performances.

Many of the consumers who reported in the October survey that they did not participate in the collection drive, stated 'small can consumption' as their reason. Of those who participated in the trial, 69% returned all the cans they had 'consumed', 10% nearly all, 13% half and 7% less than half. Assuming that 'nearly all' equals 75% and 'less than half' equals 25%, and provided the can consumption is evenly distributed among all respondents, we get the following degree of returns or retrieval rate:

$$0.6 \times (0.69 \times 1.00 + 0.10 \times 0.75 + 0.13 \times 0.5 + 0.07 \times 0.25) = 0.51$$

If the assumptions made above are correct, the 'average' retrieval rate among the year-round Varberg population would be around 50%. Disregarding the decline due to tourism and certain initial difficulties, this value would appear to be reasonable when compared with the results from the short trial period.

Concerning attitudes to alternative systems of returns, consumers were asked what they thought about the idea of introducing deposits on the cans. The answer to this question is shown in Table 10.5. It can be seen from this table that, of the summer and the permanent population, at least 75% are favourably disposed towards a deposit system.

Table 10.5

Consumer attitudes to can deposits. All values are percentages of the sample questioned

	Summer population		Year-round population	
	Can consumers	Non-consumers of cans	Can consumers	Non-consumers of cans
Positive	81	89	74	78
Negative	18	9	7	15
Don't know	1	2	19	7

Throughout the project, littering was continually surveyed in and around central Varberg. The observations furnished no evidence of a reduction in litter. There was however evidence to the contrary. Somewhat surprisingly waste bins were filled with great quantities of cans, some of which would otherwise probably have been discarded onto the ground. Quite noticeable, too, was the dwindling volume of litter around kiosks and other sales outlets. Hence, a shortage of wastebaskets at local level was remedied, while no effect could be observed on total littering within a larger area.

The reception point for the cans collected in Varberg was a local scrap depot. During the period under review, the price of aluminium can scrap was about SEK 3,600 per ton (7 öre per can) at the smelter gate. To minimise transport costs however, the can was supposed to be compacted. The return for non-compacted empty cans was therefore calculated by subtracting compacting cost and costs of transport from scrap depot to smelter from the agreed scrap price. These costs were calculated at about two öre per can, for which reason, the compensation paid out upon delivery straight to the scrap depot was five öre per can. Similar recompense was received by individual consumers or groups who wanted to sell empty cans straight to the scrap depot.

Consumers who returned cans and paper in the course of regular collections were not financially compensated. However, the storage of cans does not seem to have caused any major problems. Of those who collected cans in the home, only 3% mentioned problems with storage space and only 7% problems with offensive odours. In the opinion of 69%, the cans were taken away with sufficient frequency.

Campaign publicity in Varberg cost a total of about SEK 100,000. On top of this were the costs of auxiliary collection equipment plus the costs of collection itself. The result evaluation group, consisting of representatives from five central and local governments agencies, found it impossible to present a detailed cost-benefit analysis of the trial, but was of the opinion that the collection of aluminium cans at the same time as paper should increase the economic return of a recycling operation based on collections from households.

One by-product of the publicity which surrounded the can return trial was a sharp increase in the amount of paper collected in Varberg. However, it must be emphasised that the quantities collected in the town at the start of the trial were at a very low level when seen nationally.

RESULTS OF THE OFFICIAL INVESTIGATION

The official committee presented the results of its investigations in December 1980 and decided that a retrieval rate of 70–80% should be achieved from a system of returns. This ratio derived from a statement made by the Minister of Agriculture in Parliament in 1979, and was not linked in any immediate or obvious way to the public goals for the packaging sector discussed earlier. The reasoning which might have underlain the position taken by the Minister of Agriculture and the committee, is that energy use should be reduced. The committee certainly noted that the energy used to produce and distribute aluminium cans amounts to only between 0.1 and 0.2% of the country's total energy use but at the same time these cans are more energy-intensive than other beverage containers on the market. With a retrieval rate of 70–80%, aluminium cans would be somewhat more energy-intensive than refillable bottles which are used 30 time, and somewhat less energy-intensive than steel cans which are not recycled.

The voluntary system of returns described earlier was rejected by the committee on grounds that 'it is scarcely possible with this system to attain a recovery of 70–80% in the whole country'. Instead, the committee presented two alternatives which may be summarised as:

(a) An increase in the container levy for aluminium cans by at least 10 öre above the existing level, to take effect on 1 July 1981. Such an increase would offset the price reduction on cans which might follow because aluminium cans were about 10 öre cheaper to manufacture than the traditional tin can. The higher charge was also considered justified because of the environmental and resource-management effects that were expected to arise during 1981, when Swedish production was expected to start in the absence of any operating recycling system.

(b) An increased, refundable container levy of 70 öre for aluminium cans, to take effect on 1 July 1983. The committee wanted the refund to be set at such a level that if there were a 75% retrieval rate, the cost to the manufacturer/importer would stay at 20 öre, i.e. the old packaging charge (10 öre) plus the charge of 10 öre that the committee wanted to take effect on 1 July 1981 (according to (a)). Should the retrieval rate be less than 75%, the charge would rise by 1 öre per percentage point below the 75% level (e.g. to 20 + 25 = 45 öre with 50% retrieval). Should the retrieval rate be higher than 75%, the charge would still stay at 20 öre.

The committee's proposals ran into massive criticism on all sides and in the 'vacuum' which then arose, TEM presented the deposit proposal which was finally accepted both by the public sector representatives and by private industry.

THE DEPOSIT SOLUTION

During the autumn of 1981, the deposit solution was widely discussed by the Ministry of Agriculture, the breweries, the distributive trades and PLM. In the end they briefly agreed on the following proposals, which the Swedish Parliament adopted on 18 May 1982 (see Fig. 10.3).

1. The aluminium can is acceptable from society's viewpoint if at least 75% of it is recycled.
2. The private sector parties covenant to build up a deposit system for aluminium so that at least 75% recycling will be attained in 1985. The system will be introduced in 1983.
3. A special deposit company is to be formed for this purpose by the private business sector. The State will not own any interest in the deposit company but will assume scrutiny and control.

4. The container levy will be unchanged for the time being, but will be abolished on all containers for beer and soft drinks once the deposit system is under way. The question of introducing a differential pollution levy on beverage containers will be investigated with all appropriate speed. If adopted, the pollution levy would promote packaging systems that economise on resources and energy and that inflict minimum environmental nuisance. In principle, this means that the only types of package to be affected will be non-returnable containers. The investigative work is to be conducted quickly so that political decisions on pollution levy can be taken sometime during 1983.
5. To reduce the strains on liquidity during the initial phase, a conditional loan of Skr 50 million will be advanced to the deposit company.
6. Imported aluminium cans and domestic aluminium cans will be treated alike. When aluminium cans are imported, therefore, a deposit equal to that levied on domestic cans will be levied on the importer.
7. It is imperative for society to maintain a system of returnable glass on the beverage market. The breweries undertake to safeguard the present returnable glass system and go ahead with the work already initiated of developing a better returnable glass system.
8. Initially, the amount of deposit will be 25 öre. Should recycling fall below 75% during 1985, the deposit company will set a higher amount.
9. When selling aluminium cans to the breweries, PLM is to be paid for that deposit which accrues to each can. PLM in its turn pays this amount to the deposit company. It is assumed that an importer of empty or filled aluminium cans will pay a similar amount to the deposit company.
10. Grocery stores and other sales outlets for beer and soft drinks are to be equipped with receiving appliances (deposit machines) for empty cans or be linked to the system by some other means. As the system builds up, the principle of competitive neutrality will be applied, which means that all sales outlets can link up if they so wish. It has been estimated that between 20,000 and 30,000 appliances may have to be installed.
11. The consumer pays the deposit when he buys his beverage in an aluminium can and it will be returned when he returns the empty can. The deposit company will compensate the stores for paid-out deposits. The scrap metal which arises after the machine has compressed the returned cans will be the property of the deposit company, which will arrange to collect the scrap and transport it to a smelter. To prevent fraud, empty cans must be virtually intact to permit their being fed into the deposit machine. Cans that have been badly deformed will be recovered by separate collection systems. This may be the case, for instance, in connection with clean-up campaigns in the countryside.
12. Deposit machines, which should be manufactured in Sweden, will be owned and serviced by the deposit company. The costs of building up and operating the deposit system will be covered by the consumption paid for the scrap

metal and by the surplus that will arise because the actual recycling will not amount to 100%. The higher the degree of recycling, the lower will be this surplus. The private sector parties regard it as feasible to achieve more than 90% recycling. The surpluses that will arise, at least initially, should be set aside to meet any future losses due to variations in scrap prices and total sales volume.

FINAL COMMENTS

As noted elsewhere (see Chapter 11) two conditions should be fulfilled before using container levies:

(a) The environmental impact must have the character of an external diseconomy.
(b) The impact must be more than marginal.

It is only in respect of littering that an external diseconomy, which is not marginal, can be attributed to the system of putting beverages into containers. That is because no other type of environmental impact from the beverage container sector meets both the conditions postulated.

A deposit on one-way containers is expected to stimulate their return and cut down on their incidence as litter in the countryside. The consumer who does not return his one-way containers will incur a financial loss equivalent to the deposit. So, unlike other levy methods, the deposit method does give the consumer an incentive to return the container at the time of disposal. Even if some containers end up in the countryside, a deposit will serve as a natural incentive to pick them up. Seen from a littering viewpoint, therefore, a deposit can be expected to generate decidedly favourable effects.

CHAPTER 11

The Swedish container levy – an evaluation

KARL LIDGREN
TEM, Lunds Universitet, Sweden

INTRODUCTION [1]

In a number of countries and in some states of the USA various measures directed against the beverage container sector have been taken. In still more countries and states discussions are taking place at different levels about possible intervention. Sweden is a country where mesures have already been taken by means of the introduction of a container levy. As far as possible, an account is given here of how the problems relating to these regulations have been dealt with in Sweden during the last decade. The position is such that decisions concerning regulation gradually evolve and the possibilities of influencing the final decision are therefore not unimportant if these are taken up at the beginning of the decision-making process. The example of Sweden also provides an insight into the political moves behind measures on container levies. Nowadays this levy is 12 öre (10 öre plus V.A.T.) but only a number of coincidences prevented it from being 2–3 times higher (or perhaps even coming into existence at all).

ACTIVITIES IN PARLIAMENT

The beer can was introduced on the Swedish market in 1955 and was followed towards the end of the 1950s by the non-returnable bottle. The 1960s became the growth decade of the non-returnable container. At the same time the number of bills in the Swedish parliament dealing with non-returnable containers increased considerably. One result was that in 1967 a committee was set up, called 'The Committee for a Cleaner Society', whose terms of reference clearly show that environmental policy in the area of containers during the 1960s concerned the link between non-returnable containers and litter.

> The representations that have been made to the Crown, the bills presented to the Riksdag, and various committee reports place non-returnable containers and especially non-returnable bottles in the foreground. It is obvious

> that these phenomena affect in a decisive manner, our surroundings, both those provided by nature as well as those directly shaped by mankind, to the danger of becoming unsightly and losing any enjoyment and recreational value. In addition, the risks to health connected with being outdoors increase [2].

Among the Committee's proposals presented in 1969 was a proposal for a commission whose primary task was 'to watch consumer packaging from the point of view of littering' [3]. During the 1970s, however, political ideas about the relation between beverage containers and the environment received a more comprehensive approach. For example, the following was put forward in a bill:

> Most people perhaps disregard the destruction of raw materials for tinplate cans – raw materials which cannot be replaced. This is serious. Solely from this point of view, a prohibition of tinplate cans should be considered. We do not have the right to squander raw materials that cannot be replaced. Recovery is impossible as the cans are spread on different rubbish dumps and after some years are oxidised to rust.
>
> From the point of view of the environment both non-returnable bottles and cans of tinplate or plastic seem to be unjustifiable. The waste of money is made worse by the extra costs for waste collection. Paper packaging, for goods where the returnable bottle is not suitable, is undoubtedly the best alternative as it is easy to destroy and in addition does not use raw materials that cannot be renewed.
>
> In view of the above, it is proposed to petition the Crown for a bill aimed at prohibiting such packaging that, from the point of view of costs, stands in obvious disparity to the product sold and implies a diseconomy with irreplaceable natural resources [4].

The bill was considered by the Standing Committee on Economic Affairs which stated that 'in our opinion prohibition should be resorted to for such products that on usage or destruction give rise to environmental damage, for example, because they contain substances that are poisonous or do not decompose easily. For the remainder of the containers dealt with here, we recommend a general deposit system. This shall be constructed so that the consumer becomes disposed to return a container after use. Consideration should be given to trying to combine the deposit system with a levy system so constructed that it stimulates re-use or regeneration of the packaging. Technical developments of containers which promote re-use is thus encouraged at the same time as the consumer has some degree of choice. The use of resources is also decreased through this [5].

The dislike of non-returnable containers for beverages was not specific to any particular political party but cut across all party lines. Thus, for example,

C. H. Hermansson and others (Left Wing Communist Party) submitted 'that the Riksdag in a communication to the Crown asks for proposals to prohibit the manufacture and sale of beer and soft drinks in non-returnable containers of glass, plastic and metal within the beverage and brewing trades' [6]. This proposal was justified among other things by 'the fact that the increased use of non-returnable containers for beer and soft drinks is an enormous waste of material and energy is shown without any doubt' [7].

Thus, in 1973, a container levy was introduced as a result of the so-called Haga I agreement. The report of the Standing Committee on Taxation containing proposals for regulations concerning levies on certain beverage containers, stated that 'from the first reports on the agreement it appears that a levy of about 20 öre per container was calculated' [7]. It also said that 'it has been convincingly shown on the part of Trade and Industry that the first-mentioned levy of about 20 öre for non-returnable containers should carry with it great adjustment problems not only for the packaging and beverage manufacturers but also for subsequent trade sectors' [3] and that 'the environmental policy value of the measure in addition has been strongly questioned by the nearest authority' [3] (the National Environmental Protection Board). As a result of these and similar representations, the Standing Committee on Taxation decided to recommend a levy of 10 öre per container (both for non-returnable and returnable containers). However, the Committee pointed out that 'dealing with this matter has also suffered as the contents of the agreement in certain parts were unclear' [7].

The Committee's construction of the levy means, as it is stated, that 'as the receptacles have an average usage of 10 times in the case of the uniform levy, the actual increase in costs on the beverages that are filled into these receptacles becomes approximately one-tenth of the levy on the non-returnable containers' [3].

This misinterpretation of the real effect of the levy on returnable bottles is repeated in the Committee's proposal where it states that 'if the levy at the import of a foreign beer is 10 öre per bottle, a foreign brewer should therefore receive back 9 öre per bottle when the empty bottle is exported' [7].

According to the Committee, the reason for the proposed levy was 'partly to increase national revenues, and partly to assist in an increased use of returnable containers at the expense of non-returnable containers' [8].

However, a reservation was first made against the Committee's proposal by the Swedish Conservative Party's members of the Committee who demanded 'that the Riksdag should not take any steps because of the so-called three party agreement' as in their opinion, inter alia, 'the only visible proof of the "agreement" entered into, and whose contents are unclear, is a weakly-worded press communiqué. The "agreement" entered into is thus not of such a kind that it is fit for discussion in the Committee'.

Secondly, there were reservations by members of the Committee from the Liberal Party and the Centre Party, who also stated:

One reason why the Centre Party and the Liberal Party entered into the agreement was that according to this, it gave the consumers the possibility of avoiding price increases. The levy on non-returnable containers could be avoided by using returnable containers. A further decisive reason was that the levy on non-returnable containers has a clear purpose with regard to environmental policy.

The use of non-returnable containers implies an unnecessary extraction of scarce natural resources such as certain metals and energy. Containers can contribute to the littering problem because for one thing, a certain proportion of non-returnable containers are thrown away outdoors, when people and animals may be injured and the landscape made unsightly. Cleaning up is expensive. Non-returnable containers also contribute to an increase in the total volume of litter, which in turn leads to increased costs for refuse collection for municipalities and private individuals.

Non-returnable containers further mean that goods are usually more expensive. There are therefore strong economic reasons for stimulating a changeover from non-returnable containers to returnable containers, on condition that the latter are in practice used more than once. An increased use of returnable bottles leads to a decreased extraction of resources, less littering and lower costs to the consumers.

A levy on non-returnable containers in accordance with the Committee's opinion could possibly lead to a relative increase in the use of returnable bottles. To exert some control and produce the desired effects on the environment, the levy must be made so large that the desired changeover to returnable containers shall take place. No levy should therefore be placed on returnable containers. On the other hand, a levy on non-returnable containers should not be made so high that the reorganisation process leads to large extra costs and employment problems for the industries involved.

The Committee therefore recommends a levy fixed at 14 öre for non-returnable containers for soft drinks, beer, wine and spirit. A levy of this size should stimulate a smooth transfer to returnable containers, which allows the affected industries to adapt to changes in demand. With a levy of 14 öre, the state receives an increased income of about 140 million Skr, which is in line with the three party agreement.

As appears from the quotation from the Tax Committee's treatment, there are obvious difficulties in interpreting the three-party agreement with respect to the size of the levy, which goods were to be taxed and also the function of the levy in regard to environmental policy. This disagreement is expressed more clearly in the parliamentary debate that dealt with the Tax Committee's report on the container levy. In the Official Reports of the Riksdag on the debate, it appears that on their side the Social Democrats saw the levy initially as a means of financing certain expenditures whilst the Liberal Party and Centre Party urged the environmental policy profile of the levy. The two attitudes received sharp criticism from both Right and Left. Mr. Brundin (Conservative) stated that: 'The question is whether we really shouldn't introduce an environmental levy on the Social Democratic Party, the Centre Party and Liberal Party because they have littered the political environment through the handling of this question' [9]. The impossibility of only using one means (the container levy) in order to achieve a number of aims (the budgetary and environmental policy aims) were stressed among others by Hermansson (Left-wing Communist Party) who said:

> Where have the aspects of environmental policy now gone in the real sense of the word? The Committee has swept them under the table together with the broken glass. We are told that the effects of the new tax on environmental conservation shall be investigated quickly. Yes, at least that's said honestly; but nothing is known about the effects on environmental policy of the tax it is proposed to introduce. The environmental motive can be quite simply removed from the debate. Here it has been a case of obtaining some millions for the exchequer. The centre parties do not show the same honesty in their reservation. There the logic is not even bizarre. The Centre and Liberal Parties say that they want to increase the tax on petrol, on cars and on non-returnable containers in order to give 'the consumers the chance of avoiding price increases'. The levy on non-returnable containers can be avoided by using returnable containers, say the centre parties. How considerate towards the consumers! But what happens if the consumers really shop carefully and efficiently? Then the centre parties are left empty-handed; the exchequer doesn't get the millions that were the basis for the new tax [10].

The proposal of the Standing Committee on Taxation for a levy of 10 öre on all beverage containers (excluding those used for milk) was approved by 138 votes to 106 for the Liberal and Centre Party's proposal for a 14 öre levy on non-returnable containers only. The levy is paid for all new production in Sweden and for all imports. Furthermore, the levy is only paid for containers larger than 20 cl and smaller than 3 litres.

The official enquiry into the costs of environmental conservation received the task of investigating the effects of the container levy. In the report of the enquiry [11], the result is stated to be that the levy had not had any great influence on the costs, employment or environment. A limitation to environmental effects alone does not change the picture of a certain but comparatively insignificant effect. However, a clear movement towards containers of less than 20 cl was observed as these were not subjected to any levy.

In 1975, it was time for a new Haga II round, which included an agreement for increased beverage container levies. The agreement meant an increase of 5 öre on returnable bottles and 20 öre on non-returnable containers. This agreement however was broken directly by the Minister of Finance, who recommended a levy of 20 öre on metal containers and 10 öre on other containers (both non-returnable and returnable) [12]. The Committee on Taxation in its turn changed the proposal in the bill and alternatively recommended no increase in the container levy. In practice the Committee's proposal was carried out.

One important reason why the Committee on Taxation did not consider itself able to propose any change in the container levy was that the Commission on the costs of environmental conservation was investigating the need for, and construction of, a future packaging levy. The Commission states in its report, [10]:

> On newly manufactured containers, intended for ready-to-use beverages (excluding milk), whose volume exceeds 20 cl, there is at present a container levy of 10 öre plus V.A.T. The introduction of this levy was initially motivated by the need for a special revenue for the state as a contribution towards financing the price freeze on certain basic foods. As the subvention is still in force the Commission formulated its further proposals on the assumption that the levy will be retained until further notice.
>
> The Commission considers it necessary, as mentioned earlier, that special financial resources should be created for measures against littering. The increase and expansion of the levy suggested by the Commission has as its purpose to constitute the main basis for this. This is, amongst other things, motivated by the desire expressed in the enquiry's terms of reference to let the polluter pay the costs he causes if possible. Applying the principle that the polluter shall pay, demands that the impact of littering on the environment can be determined, and that the polluter can be identified. Neither of these two basic demands however can be fulfilled here. All that is known is that certain people litter and that this causes certain nuisances such as injuries, aesthetic stress, etc. Of course, a somewhat less ambitious but perhaps more realistic approach could be made. This may imply that society requires that the costs for carrying out activities against littering shall be paid by those who use potential litter objects. As it is not known who is responsible for littering, one expedient may be to impose a levy on the whole of the sector that produces potential litter objects. Containers of various kinds, not solely beverage containers, constitute however, the dominant proportion of litter in litter counts. The Commission has therefore considered the possibility of introducing a general container levy. For the stated reasons, however, such a levy has already been rejected.
>
> If certain litter objects were to be subjected to a levy, there are undoubtedly reasons for selecting the assortment of beverages subjected to a levy today.

It is also worth noting that the aim of making the polluter pay, is not completely fulfilled. The construction fulfills only the demand that certain individuals – the polluter as well as the non-polluter – shall pay certain costs. But the Commission considers that there are overwhelming reasons for suggesting that the present levy construction, that is to say the levy on new production, be retained. It therefore proposes that the present levy be increased from 10 öre to 15 öre.

Clearly then, there have been many political moves concerning beverage containers and their environmental impact, a situation not confined to Sweden. The position still appears to be that there is no clear *political* unity about whether there should be a container levy, why there should be a container levy, and, if there were a levy, how large should this levy be.

One way of trying to evaluate the container levy in Sweden is to decide the problem that the levy is supposed, at least partly, to eliminate. To identify the problem, two conditions should be fulfilled. These are (a) the environmental impact must have the character of an external diseconomy, and (b) the impact must be more than marginal.

Before going through the conditions, one shall bear in mind the following rule, accepted by professional economists: for a successful policy you should use at least as many measures as there are aims!

Condition 1: External Diseconomy

An external diseconomy arises when the production of one or more firms and/or the consumption of one or more individuals affects the production possibilities for other firms and/or the living standard for other individuals, with no payment being made for this impact by those who cause it. One example of an external diseconomy is a plant's emission of smoke into the air which adversely affects the persons living in the surrounding area. The consumption of a bottle or can of beer on the bathing beach will directly affect the wellbeing of other individuals if the container, after use, is discarded on the beach where, apart from its unsightly appearance, it may give rise to accidental injury.

Now how do different types of environmental impact meet the criterion that a diseconomy must be shown to exist?

Starting with the link between natural resources and packaging, it is clear that market-determined prices are paid for the use of natural resources, so no external diseconomies will arise. Certainly, the actual use of different natural resources may cause pollution of water, air and land but this aspect of natural resource consumption is considered under the heading of impact on the external environment.

However, many people feel that we are going through a natural resource crisis and that any public policy aimed at reducing the use of natural resources is worth considering. Two questions now present themselves. Are we going through a natural resource crisis? If so, how can this be attacked?

To the first question there is no direct answer, but if natural resources are to be saved they must be made more expensive no matter what their application. Obviously a national 'natural resource tax' will also affect the packaging sector. But it permits resource savings to be made at a rate and in a sequence which the private decision-makers (consumers, corporate executives, etc.) think best, at the same time as the overall result is brought into line with the community's objectives due to variations in the tax bite.

When it comes to impact on the external environment in the form of water, air and land pollution, it is clear that external diseconomies are involved. Water pollution is typically exemplified by effluents of chemicals from pulp and paper mills and of detergent lye from breweries, notably resulting in eutrophication of lakes and waterways. As a result, the water is less able to sustain other activities such as recreation or as a source of mains water, but this is not reflected in the polluter's accounts as a cost of doing business.

As another example, the manufacture of wood pulp, tinplate and glass is accompanied by emissions of sulphur dioxide, which in turn not only acidifies land and water but also inflicts corrosion damage. Again, the polluter does not bear the economic consequences of his conduct. The fact that the factories pay for purification plants to treat one part of their waste does not vitiate this observation.

Lastly, when we consider land pollution, the impact covers a broad spectrum ranging from changes of landscape due to gravel pits, etc. at one extreme, to littering at the other. Even if littering is outlawed, the polluter escapes payment liability provided he is not caught and brought to book.

When consumers pay the municipality for collecting and disposing of their refuse, they are effecting compensation for the resources which are used up in the form of labour, fuel etc. However, no charge is made for the pollution of water, air and land which arises in connection with handling this waste.

With a given technology for waste management (a concept that includes recovery and recycling), different containers may be more or less expensive to treat. This in turn can be made to underpin any intervention in the packaging sector intended to create opportunities, already in the project stage, for keeping down the costs of waste management. The course of events can be steered both by banning certain products and by the imposition of refuse collection charges. Under the scale of charges now in force, a specific amount is payable per unit of volume irrespective of the quality of the waste, even though different grades of waste may give rise to different treatment costs. However, since it is virtually impossible in the real world to identify who is discarding what type of waste, a more feasible alternative would be to put the refuse collection charge directly on the product. Unlike today's refuse collection charges, which are paid at the time of disposal, 'tomorrow's' charges would be paid at the time of purchase and be made contingent on the quality of the end waste represented by the emptied container. However, such a procedure postulates real differences between containers in their waste management costs plus advance knowledge of how large these costs are. Nevertheless, it is hardly correct to speak here of external diseconomies inasmuch as the total costs are paid de facto. It would be more proper to speak instead of imperfections in the pricing system itself which, if they are seen as troublesome, can be set aright by imposing statutory controls on package design or by imposing charges.

Summing up, it seems that external diseconomies can only be established in respect of water, land and air pollution and that these can justify direct intervention against the packaging sector.

Condition 2: Marginal Impact

The first condition – that an external diseconomy shall be shown to exist – was met by the external environmental impact which the packaging sector causes. However, yet another condition must be met if direct intervention against the packaging sector are to be meaningful. Environmental impact from the packaging sector must be more than marginal compared with total environmental impact of the same kind which takes place in the community at large. Several examples can be selected to show why this condition should also be met.

Where containers are distributed by trucks, the packaging sector's impact on the external environment takes the form of exhaust gases and noise. The manufacture of tinplate intended for containers is accompanied by discharges which affect both water and air. Paper, a material used in many containers, causes different types of environmental impact during production. So does the manufacture of glass.

These examples illustrate that there are different types of environmental impact from the packaging sector. Collectively they can be termed 'external diseconomies' but the interesting aspect here is that such environmental disturbances arise from many different activities in the community and that the packaging sector's contribution is usually marginal. All road transportation generates fumes and noise. All iron and steel mills pollute water and air. All pulp and paper mills discharge organic effluents. Thus, if measures aimed at reducing such environmental impacts are to be administratively manageable, and are to lead to genuine reductions when seen overall, they must be generally applied. So all motor vehicles must conform to more stringent rules on exhaust gases and noise regardless of what they transport. Tighter rules on water and air pollution must be enforced against all pulp and paper mills regardless of what their products are used for. And the same, of course, should also go for the iron and steel mills.

Confining considerations to the beverage containers, around which most of the packaging–versus–environment debate has revolved, the following observations can be made.

It is only in respect of littering (according to surveys 20% of the litter strewn along Swedish roads and highways consists of beverage containers including caps) that an external diseconomy can be attributed to the system of putting beverages in containers. The effect is not marginal. This diseconomy is recognised because no other type of environmental impact from the beverage container sector meets both of the conditions postulated.

CONTAINER LEVIES FOR PREVENTING LITTERING – CALCULATION METHODS

In the theory of negative external effects, levies are often suggested as a useful control measure for influencing effect-generating activity so that its level becomes optimal in regard to the national economy. The advantages of using levies as control measures in preference to different regulatory measures have been stated as:

(a) Fixing levies requires the availability of less information.
(b) Levies lead to specific targets being reached at a lower cost to the community.
(c) Levies mean a continuous pressure on the polluter to look for a less 'environment-wasting' technique.
(d) Levies mean the polluter paying for the residual discharge. This in turn fulfils the demand that the polluter pay the environmental cost.

In general, levies can be recommended as control measures to deal with some national economic environmental problems. We have therefore considered whether this also applies to levies on containers in order to deal with the littering problem.

An external effect is a type of pollution for which no payment is made by the person causing the pollution. It follows therefore that one way of eliminating negative external effects would be to let those who cause them stand for the costs also; in this case, by means of a levy on the beverage container. Such a procedure requires however that inconveniences (costs) can be estimated and charged to the person who creates them.

Under such a system, a discarded container that causes small inconveniences should be subjected to a smaller levy than a container of the same type, which causes a greater inconvenience. In practice such a levy is impossible to impose simply because it is impossible to measure the inconvenience of every discarded container. Furthermore, it is not possible afterwards to extract a levy from the person who discarded the container. Instead the construction of a levy must be carried out in such a manner that a levy is equitably charged to the consumer, the potential litterer, when he purchases the beverage container. The difficulty of differentiating the levy for a certain type of container does not of course obviate the possibility of differentiating the levy for different container types. One consequence of this, however, is that, regardless of how high the levy is set, most consumers pay a levy which does not correspond to the inconvenience resulting from the manner in which they dispose of the empty container.

The statement that the use of beverage containers gives rise to external effects can be taken as the point of procedure for a discussion on how a levy should be constructed. What is of interest here is the costs to be used as a basis for calculating levies and over which container costs are to be spread.

An initial method assumes that all beverage containers may finally be discarded outdoors and should therefore be dealt with accordingly. Hence all beverage containers bear a levy corresponding to an average value of the negative external effects produced by a container discarded outdoors. This is referred to as the high levy method.

An alternative is to arrange that all beverage containers pay the value of the external effects that only certain beverage containers cause. In this way, the total activity that the consumption of beverage containers causes will bear all 'environmental costs' that this activity gives rise to. The probability that a container will be disposed of outdoors together with the value of the external effects, constitutes the basis of calculation for the levy. This method is referred to as the low levy method.

The beverage containers discarded outdoors lead to different types of external effects such as injuries to people and animals, aesthetic effects, clean-up costs, etc. These effects can be of varying degrees of difficulty to identify and evaluate; this can be taken as an excuse to disregard some of them completely. One way of doing this is to say that the only important external effects are injuries to people and animals. Another way is just to count the cleaning-up costs [13].

The extraction of container levies can in principle occur either in the shape of levies on the production of new containers, or as a levy on each beverage container sold in some later sector. The difference between these levy constructions affects only returnable bottles, as the first construction leads to an increase in the deposit, which the other does not. The consumer's possibility of avoiding the levy, when the second levy construction is utilized, consists only of decreasing his consumption [14] whilst the first levy construction also provides the possibility of changing over to returnable bottles that are returned in order to avoid the levy. Consider the levy construction that calls for levies on the production of new containers.

If the high levy method is used, the total revenue will be greater than the value of the total external effects. Through this method, the consumer who purchases a container subjected to a levy will thus be treated 'unfairly' if he discards the empty container in some dustbin whilst he will be more 'justly' treated if he discards the container outdoors. This situation can have the effect that consumers will to an increased extent discard empty beverage containers in the open as they consider they have paid for this right in spite of the intention to limit littering. In addition, the high levy method has two further marked weaknesses. First it has a controlling effect only at the time of purchase and not a disposal as there still remains no difference between discarding a container that bears a levy and one that does not. The exception is the returnable bottle where the levy is paid at the production of returnable bottles. Such a procedure of course increases the deposit. A levy on non-returnable containers does not increase the potential litterer's cost in his calculation on a decision to litter and so differs, for example, from a deposit. Second, the levy causes certain consumers to change to returnable bottles although they earlier threw the empty non-returnable container into some kind of dustbin. They have thus a decreased option although there is no reason for this.

In Sweden clean-up costs have been calculated at an average of 50 öre per discarded container. If the other inconveniences are valued at the equivalent amount, the levy per container should be 1 Skr. It is unlikely that non-returnable containers could stand such a levy and so the high levy method is equivalent to a prohibition. In Norway, where there is a levy of 0.80 Nkr for a non-returnable container for beer or soft drinks, the use of non-returnable containers (cans) is almost negligible.

In contrast, the low levy method controls neither at the purchasing point, nor, unlike the deposit method, at the time of disposal. If the inconveniences are presumably 1 Skr and 1% of the containers are thrown away, the levy becomes 0.01 Skr per container. A doubling of the inconveniences to 2 Skr and the number thrown away to 2% makes the levy 0.04 Skr. Even if the higher levy of 0.04 Skr were used, it would adjust the price of container plus beverage by only a small percentage. With an inflation rate of perhaps 10% per annum, the introduction of such a levy would hardly be noticed in the general price increases and would be almost meaningless if it is intended to have a controlling effect. It is worth noting that both Ahrell and Helgesen consider they could observe that consumers' knowledge of prices was bad, so the influence on the consumer of small price variations are thus further lessened. For such environmental levies to be relevant, the littering situation must be noticeably improved, otherwise the levy is nothing more than a pure tax on consumption and as such can just as well, and perhaps even better, be obtained as a general tax. Some critics may however consider that it is 'fair' that the beverage container sector, as a collective, ought to pay for the external effects, such as littering, that it causes [10]. Against this it can be said that the non-litterer is also hit by this 'fair' levy.

EVALUATION OF THE SWEDISH LEVY

As mentioned earlier, a container levy was introduced in Sweden in 1973 on the new production of containers greater than 20 cl but less than 3 litres, intended for ready-to-consume beverages, excluding milk. The levy is 0.10 Skr plus V.A.T.; that is, approximately 0.12 Skr. This levy is in line with that determined by the low levy method and so it is worth considering its effects more closely.

The effects of the levy were examined for the years 1972 and 1973. It was found that the trippage of returnable bottles increased and the proportion of non-returnable bottles in rubbish decreased. The market share of returnable bottles increased and, since overall consumption was unchanged, this implies that non-returnable containers lost some of their market share. There are, however, many explanations of this changeover to returnable bottles in addition to the fact that the levy signified a relative price increase for beverages packed in non-returnable containers. First, returnable bottles were once again introduced into many shops that had previously only stocked non-returnable containers. This increased the consumer's freedom of choice, which in itself could have led to a choice of returnable bottles. Secondly, the possibilities for returning returnable bottles were improved and the consumer's total cost for using returnable bottles was decreased because the return process was made less troublesome. Thirdly, returnable bottles in crates came into use as special offers and often the consumer price was fixed below the price paid by the shop to the brewers. Finally, the introduction of the levy and the accompanying discussions which implied propaganda against non-returnable containers and for returnable bottles, received further currency. This in turn may have led to changes in preference, to the advantage of returnable bottles. Thus with the levy, there were at least four other factors which could have contributed to an increased transition to returnable bottles.

In a mixed system containing both returnable and non-returnable beverage containers, combined with a relatively high deposit on returnable bottles, very few returnable bottles are discarded outdoors. Thus it can be seen that littering with beverage containers occurs almost soley with non-returnable containers in such a system.

Did the levy result in less littering? Using Autumn 1972 as a basis for measurement the relative changes of the refuse level can be described. As Table 11.1 shows, the total quantity of rubbish has decreased since 1972 whilst the number of beverage containers has increased, except for a decline in 1973.

Table 11.1

Relative changes in rubbish levels and different container types sold. The values for autumn 1972 are taken as the basis (= 100). Source: National Swedish Road Administration (1976)

Year	Whole rubbish level	Total beverage containers	Non-returnable bottles	Beer cans	Paper/ plastics	Returnable bottles
1972	100	100	100	100	100	100
1973	77	85	70	86	110	60
1974	87	110	88	130	98	65
1975	82	110	73	136	104	45

The number of non-returnable containers consumed decreased from 820 million to 785 million between 1972 and 1973, (approximately 4%). The levy can be one of a number of factors that contributed to this decrease. If we follow the assumption of a constant relation between the number of non-returnable containers consumed and the number discarded outdoors, this increase in overall consumption should be accompanied by a decrease of 4% in littering with non-returnable containers between 1972 and 1973.

But Tables 11.1 and 11.2 shows that the number of non-returnable containers decreased by as much as 12% at the places where the rubbish count was carried out. One interpretation of this could be that among the consumers who gave up non-returnable containers there were more litterers than the average for the remaining consumption. However, the situation was that the total quantity of rubbish also decreased by 23%. This is considerably more than was the case for non-returnable containers, although these were the only refuse components subjected to an 'environmental levy'. The explanation of this general fall in the rubbish level probably has some connection with the fact that in 1973 there was a nationwide anti-litter campaign. Because of this, the effects of the container levy became very difficult to assess. It is clear, however, that the relative proportion of non-returnable containers increased from 9.0 to 9.9%, that is to say by 10% between 1972 and 1973 according to Table 11.3. One interpretation of this is that if the effects of the campaign are disregarded, the consumption decrease of 4% (possibly caused in part by the levy) does not necessarily imply any decreased littering. Obviously, criticism can be directed against this interpretation, and the effect of the levy should therefore be discussed using data only for the years 1972 to 1974 when no anti-litter campaign was in progress. It should be noted that littering with non-returnable containers increased during 1974 to a level clearly above that of 1972, although the number of non-returnable containers decreased further between 1973 and 1974 from 785 million to 780 million. Thus comparisons between 1972 and 1974 show that the consumption of non-returnable beverage containers decreased at the same time as their contribution to litter increased.

Table 11.2

Different types of beverage containers present in litter in 1972–1975 in Sweden, expressed as a percentage of the total number of beverage containers in litter

Year	Wine and spirit bottles	Returnable bottles	Non-returnable bottles	Beer cans	Paper/ plastic containers	Total
1972	3.4	8.0	16.4	51.4	20.9	100.0
1973	3.0	5.6	13.5	51.9	26.0	100.0
1974	2.7	4.7	12.9	61.0	18.8	100.0
1975	2.1	3.2	10.7	63.8	20.1	100.0

This suggests that the 'environmental levy' introduced in 1973, which meant a price increase of approximately 0.14 Skr per container, had little to do with the consumer's choice of beverage container type and hardly any effect at all, on his littering behaviour. Instead the effects of the container levy are found in the producer sector. The container producers rapidly changed the size of all containers that earlier held 21 cl to 19 cl. Through this alteration the container levy was avoided. The beverage producers felt an increased uncertainty in the selection of a container with the introduction of the container levy. This gave rise to opposition against the introduction of new packaging systems whose containers were not taxed, and a desire to concentrate the use of non-returnable containers to one type of container. The decreased use of non-returnable bottles that had started already in 1970 was thus accelerated markedly.

Table 11.3

Percentage volume composition of litter in Sweden in 1972–1975. Analyses were made in the autumn

Component	1972	1973	1974	1975
Bottles[a]				
wine and spirit bottles	0.5	0.5	0.4	0.4
returnable bottles	1.2	0.8	0.7	0.7
non-returnable bottles	2.4	2.2	2.1	2.1
Beer cans	4.3	4.8	6.4	7.2
Paper/plastic beverage containers	1.8	2.4	2.0	2.2[b]
Total containers	10.2	10.7	11.6	12.6
Other bottles	0.2	0.2	0.2	0.3
Other paper/plastic combinations	2.7	2.6	3.4	2.7
Pure plastics	10.2	10.6	12.3	11.9
Pure cardboard/paper	61.2	61.6	59.3	61.1
Preserve and other cans	0.4	0.4	0.3	–
Capsules, lids	8.5	7.7	8.1	–
Other metal objects	1.5	1.4	1.2	7.2[c]
Textiles, rubber	2.0	2.4	1.9	2.2
Other components	3.3	2.6	1.8	1.9
Totals	100.0	100.0	100.0	100.0

[a]Broken glass is assumed to have the same proportion for different beverage containers and other bottles as whole bottles.

[b]Of this, 0.4% paper/plastic containers for milk.

[c]Total all metals.

With the exception of reasonable deposits on returnable bottles, it thus appears that the control of non-returnable containers by the use of levies as recommended by economists, is not especially effective. It is therefore a pity that levies are used so industriously in this particular area and not within other 'environmental areas' where they may function more effectively as control measures.

REFERENCES AND NOTES

[1] This paper is based primarily on knowledge and experience gained as an expert in several official committees dealing with different 'container questions'.

[2] Terms of reference of the 'Committee for a Cleaner Society' according to the minutes of the Council of State concerning social matters 1967–03–03.

[3] SOU (1969:18), p. 146 (Sweden).

[4] Bill (1972:1363) (Sweden).

[5] The Standing Committee on Economic Affairs (1972:51) (Sweden).

[6] Bill (1972:91) (Sweden).

[7] The Standing Committee on Taxation (1973:3), p. 1 (Sweden).

[8] The Standing Committee on Taxation (1973:3), p. 3 (Sweden).

[9] Official Reports of the Riksdag No. 24, p. 26 (Sweden).

[10] SOU (1976:35) (Sweden).

[11] SOU (1974:44). 'The effects of the container levy' (Sweden).

[12] Bill (1975:92) (Sweden).

[13] The US Environmental Protection Agency in its Second Report to Congress (1974) put forward the same idea. The suggested levy was 0.5 cents.

[14] If the levy is the same, regardless of container volume, the possibility arises of buying larger containers.

CHAPTER 12

Evaluation of one-way cartons and returnable bottles for milk – economic aspects and ecological background

HEINRICH WIETBRAUK
Federal Dairy Research Center, Institute of Economics and Market Research of Food Processing, Kiel, Federal Republic of Germany

INTRODUCTION

Against the background of increasing waste from packages in household and industry and a growing awareness of pollution, it was decided in 1979 to conduct a market study on the feasibility of reintroducing returnable packages for pasteurised milk. The study was performed in the Köln area from June 1980 to May 1981. Consumer acceptance was measured in terms of market share and price recordings, consumer attitude was determined by polls and consumer behaviour towards the return of multi-trip bottles was observed. In a comparison of one-way cartons and returnable bottles, the ecological balance is determined by the circulation frequency. Also the economic comparison of both systems could be influenced by the customer's behaviour regarding the bottle return.

This chapter attempts to show the relevant aspects for the assessment of alternative packing systems in individual dairy enterprises.

In order to make an overall assessment, the following factors are important:

– consumer's acceptance
– circulation frequency
– system costs
– ecological aspects

CONSUMER'S ACCEPTANCE

The relevant parameters for consumer's acceptance are market share and retail prices in the segments of the liquid milk market. Figure 12.1 shows the market share for different types of packages as measured in the pre-test period and the main-test period of the Köln market study.

Based on total milk sales, measured in litres, the 1-litre returnable bottles reached a market share of 11%, divided into 9% for the glass bottle and 2% for the plastic bottle. Comparing the pre-test period and the main-test period, carton-packed, pasteurised, whole-milk remained at a market share of 45%, while plastic bags and other packages lost 3% of their market share. The total market share of pasteurised whole-milk expanded by 7% (from 54% to 61%) due to the introduction of returnable bottles. In particular, the market share of low-fat milk decreased – for example, the share of plastic bags fell from 18% to 13%. The market share of UHT-milk remained almost constant.

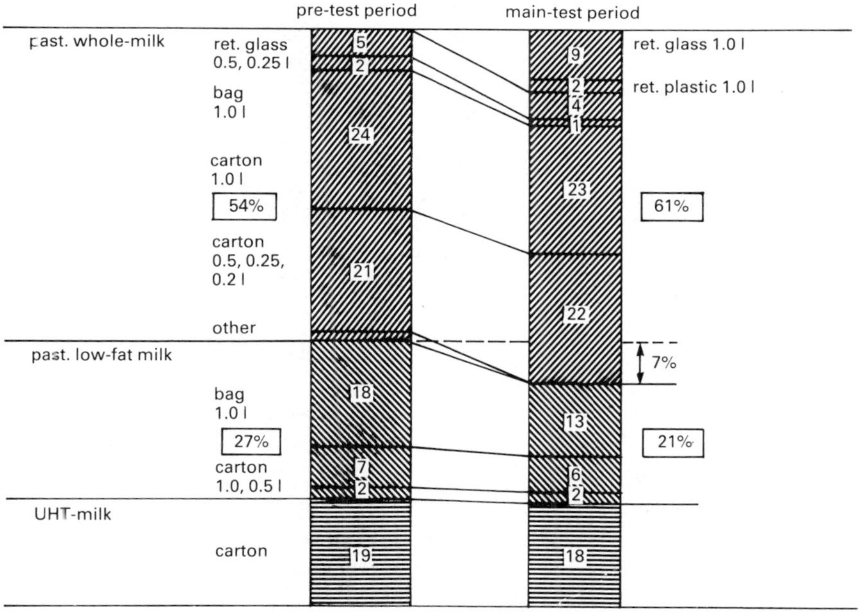

Fig. 12.1 – Percentage share of the retail market in 1980 and 1981 for milk in different types of containers. (Source: reference [1].)

Figure 12.2 shows the changes in the market segment of pasteurised whole-milk in 1-litre packages, where new packaging alternatives were offered. In this sector, returnable bottles reached a market share of almost one third, the glass bottle reached 25% and the plastic bottle 6%. Corresponding losses were observed for other packaging systems.

Consider now the second parameter which determines consumer's acceptance; the consumer price. Considerable price differences were observed between the 60 shops involved in the study. These differences concern not only the general

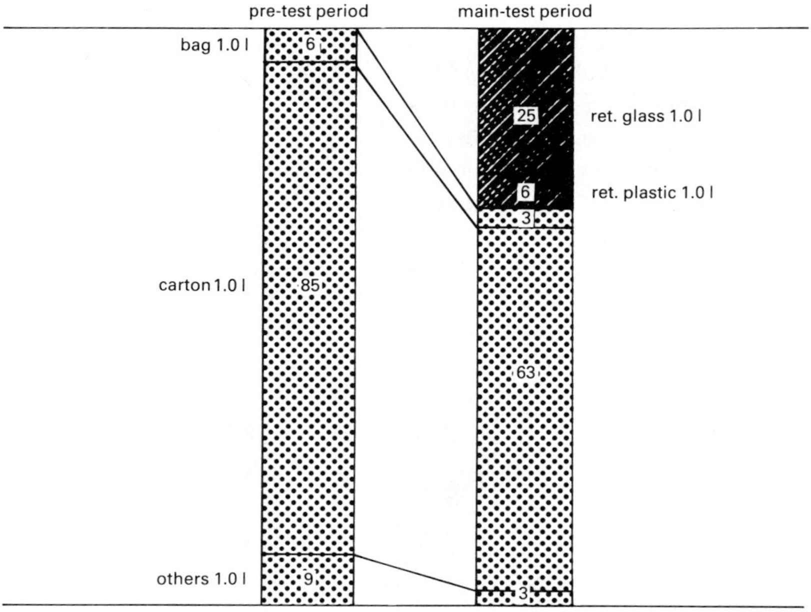

Fig. 12.2 – Percentage market share of the retail market for pasteurised whole milk in 1-litre containers of different types for 1980 and 1981. (Source: reference [1].)

price level, but also the price structure of the milk products examined. Table 12.1 shows that glass bottles were sold at an average price of 1.19 DM excluding a deposit of 0.30 DM/unit. This meant an additional amount of 6 Pf/unit compared with the carton package. The plastic bottle was sold at a price of 1.23 DM, which was even 10 Pf/unit higher than the price for comparable carton packages. Since there was a high rate of replacement between pasteurised low-fat milk in plastic bags and returnable packages this fact must be considered too. Those customers who bought glass bottles instead of low fat milk in plastic bags paid 34 Pf/unit more.

CIRCULATION FREQUENCY

The ecological and the economic evaluation of returnable packaging systems are mainly influenced by the number of trips made by the bottles. The general circulation frequency is influenced by a number of factors with each having a different relative importance according to the type of pack.

Technical factors such as breakage during production, have little effect on the circulation frequency of returnable bottles for milk. Consumer behaviour regarding return of empty bottles plays a predominant role. This behaviour is dependent on the customer's knowledge that the bottle is a returnable package.

Table 12.1

Consumer prices in the market trials (pfenning per unit). Source: G f K Nurnberg (Ref. [1])

Commodity	Price during test period (pf/unit)
Pasteurised whole milk	
in 1 litre returnable glass	119
in 1 litre returnable plastic	123
in 0.5 litre returnable glass	78
in 1 litre bags	113
in 1 litre cartons	113
in 0.5 litre cartons	80
Pasteurised full cream milk	
in 1 litre bags	85
in 1 litre cartons	95
in 0.5 litre cartons	77
HT milk (in cartons)	
1 litre whole milk	106
0.5 litre whole milk	77
1 litre full cream milk	83
1 litre entrahmte	76

Additionally, it is influenced by the size of the deposit, consumer convenience, the size of the garbage containers at home, and other alternatives for the disposal of empty bottles, e.g. recycling containers. Consumer behaviour is also influenced by the number of possible ways of returning bottles offered by the retail trade and a general knowledge about the importance and purpose of returnable packaging systems.

The circulation frequency is defined as the average number of fillings of the same bottle. It can be calculated by the formula:

$$\text{Circulation frequency} = \frac{100}{\%\ \text{bottles lost}}$$

The calculation is more complicated in practice because it is difficult to determine the exact losses. This is due to a number of interfering factors, such as the question of whether a bottle has to be included in loss quota or whether it is part of the system-related stock of bottles in households or in trade. Moreover, particularly when multi-trip bottle systems are newly introduced, a time-lag occurs while the system stocks up with new bottles.

Three different methods have been used to calculate circulation frequency as exactly as possible [2], see Fig. 12.3. These are (a) stock calculation, (b) production calculation, and (c) deposit calculation including own losses.

The different results obtained by these three methods must be coordinated and interpreted taking account of the different errors involved. As a result, it was found that the circulation frequency of glass bottles was 11 trips and of plastic bottles, 8 trips.

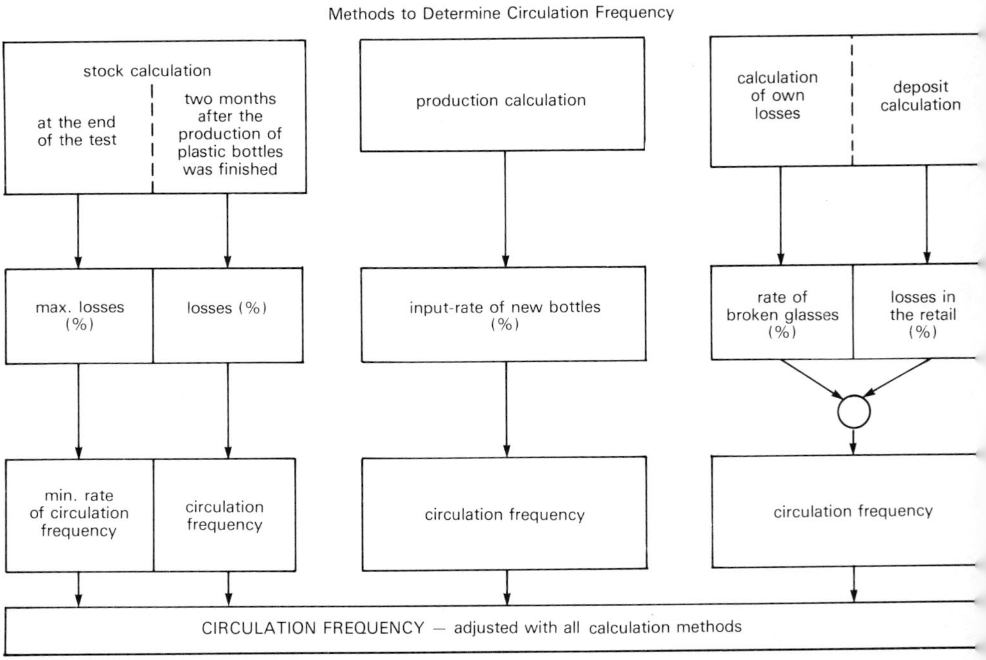

Fig. 12.3 – Graphical representation of the principal methods used to estimate bottle circulation frequency.

SYSTEM COSTS

Cost comparisons were made using contributions involved in production, distribution and retail. They are based on a model which, as a special form of part cost accounting, considers only those cost elements which are typical for the system. Indifferent system cost elements, such as pre-treatment of milk in the diary plant or display in food retail outlets are omitted. Prices at 1 April 1981 were used in the calculations [2].

Model conditions

In the filling area, two plants of the Tetrabrik-system, representing one-trip carton packages, were compared with a cleaning, filling and sealing plant for glass and plastic bottles, which allows a nominal output of 6,000 bottles/hour.

Distribution includes dispatch, transport, delivery, return and accounting of empty bottles. The model calculation is based on a delivery tour for fresh milk products. The milk products, pre-selected for each delivery tour, are taken to the customer on trucks of payload 7.5 tons. The weight of the products is the limiting transportation factor. With an average loading capacity of 3,150 kg/tour, 25 customers can be supplied with one-way cartons, while it is possible to supply only 19 customers per tour with milk filled in the heavier glass bottles.

For milk sold through food retailers, the costs of handling filled and empty packages were studied using a supermarket with average sales of 200 litres milk/day. The calculation is based on manual reception of the empty bottles with immediate pre-sorting and cash payment of deposit.

Total system costs

Figure 12.4 shows the system costs incurred in the different functional areas [2]. With a yearly output of 12.87 million units, the production costs amount to 14.2 Pf/unit for the one-way carton. This is 5.1 Pf/unit more expensive than the returnable glass bottle with 11 trips, but still 0.7 Pf/unit cheaper than the plastic bottle with 8 trips.

In distribution, the relative costs of cartons and glass bottles is reversed. Higher package weights and additional costs for the reception of empty bottles produce glass bottle costs 3.7 Pf/unit more, provided the retail outlet takes 200 units per delivery. At 8.5 Pf/unit the plastic bottle holds an intermediate position but lies closer to the glass bottle.

The sum of these costs gives the cost for retail sale. Under defined conditions, the costs for the returnable glass bottle are 1.4 Pf/unit lower than the costs for the one-way carton. For the plastic bottle, retail costs amount to 23.4 Pf/unit; this means extra costs of almost 3 Pf/unit.

The packaging systems differ significantly in retail handling costs. For the one-way system, handling costs for the filled packages amount to 4.3 Pf/unit. For the returnable system, the reception of empty bottles is an additional cost factor. For manual reception with a cash deposit system, total retail handling costs of 12.0 Pf/unit arise. Because of the lower number of returned plastic bottles and lower stock losses, the costs for plastic bottles are 0.2 Pf/unit lower than the costs for the glass bottle.

The costs incurred in food retailing influence the result of the total system comparision significantly. The cost advantage of the returnable glass bottle excluding retailing, changes into a cost disadvantage amounting to 6.3 Pf/unit. The extra costs for the plastic bottle amount to 10.7 Pf/unit compared with the one-way carton. In all three areas investigated, the plastic bottle is the most expensive package.

Figure 12.5 shows the share of the different cost types of the total system costs. Packaging materials for the one-way carton amount to 45% of the total, whereas for the multi-trip glass bottle they amount to only 10%. On the other

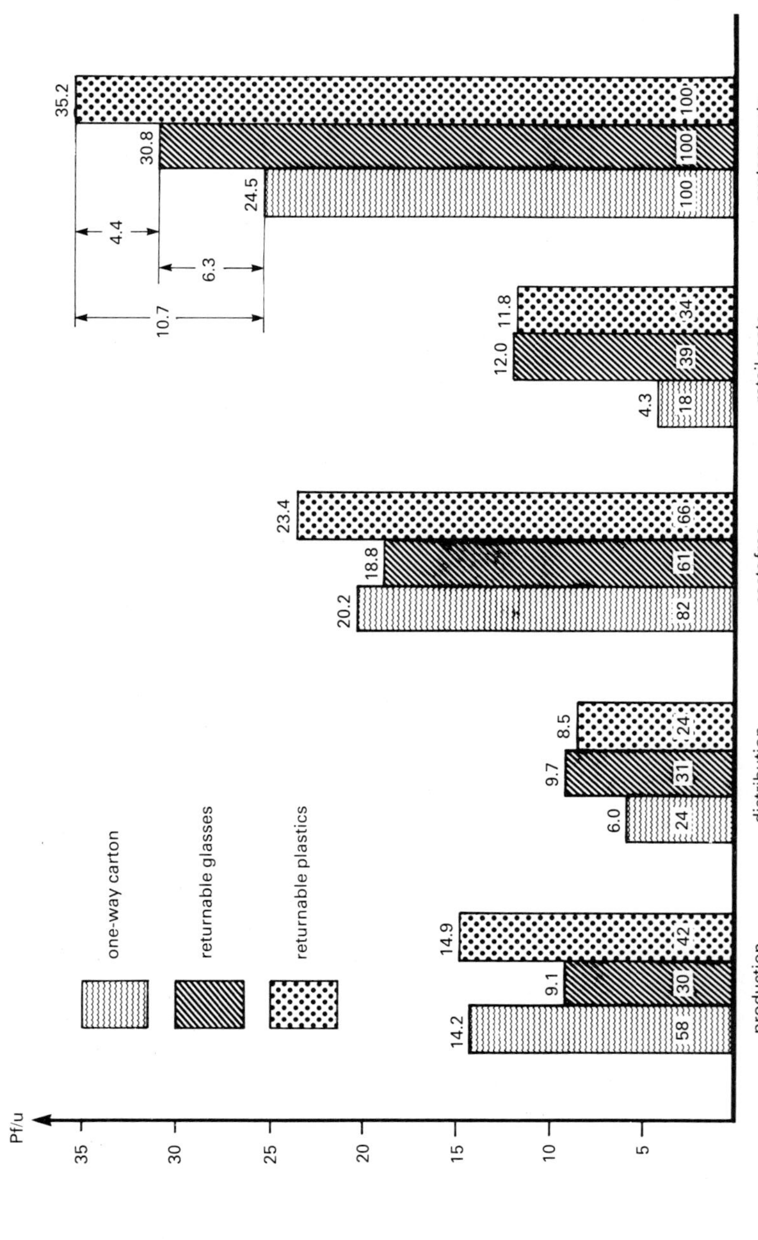

Fig. 12.4 – Distribution of costs amongst the different functions within the milk packaging system for the different types of milk packages.

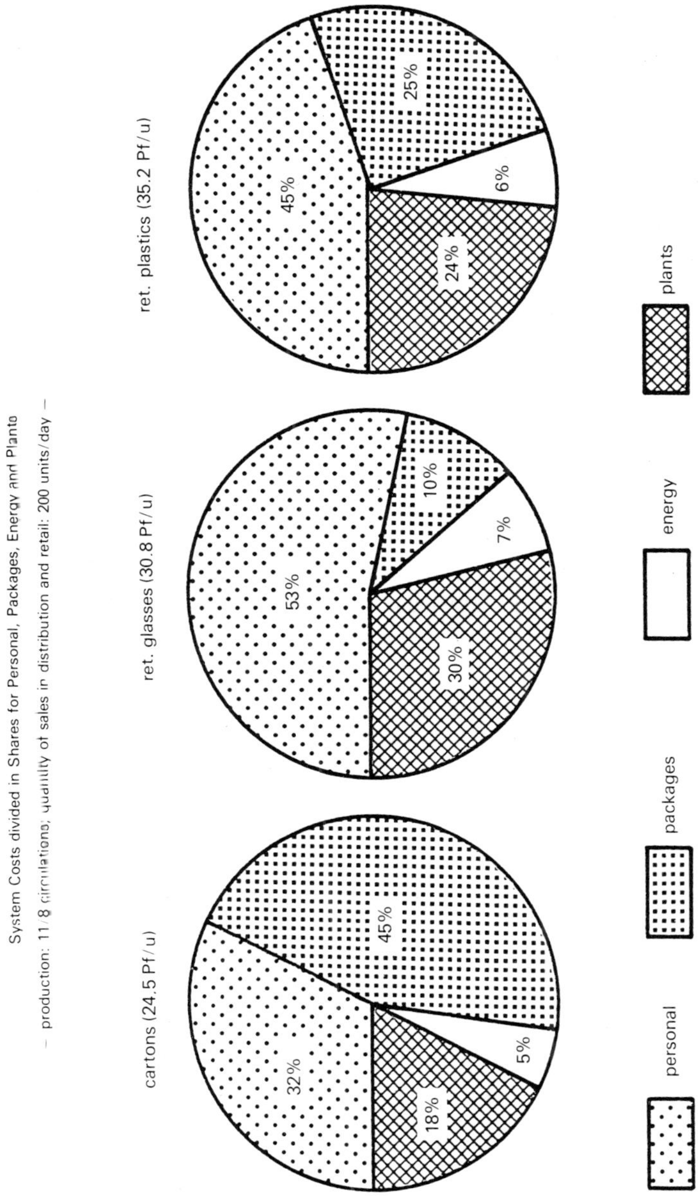

Fig. 12.5 – Share of different cost types to the total system costs.

hand, labour costs for the glass bottle account for 53%, and are even greater than packaging material costs for the one-way carton. Labour costs at 32% are lowest for cartons. For all cost types, the returnable plastic bottle occupies an intermediate position between the other packages.

Under the same conditions as in the previous illustrations, Fig. 12.6 shows the system costs as a function of circulation frequency for returnable bottles. It can be seen that with any circulation frequency, the costs for multi-trip packages are considerably higher than the costs for the one-way package. Even with an unlimited number of trips, there exists a considerable system difference.

Using the circulation frequencies of 11 trips for glass bottles and 8 trips for plastic bottles, the differences in unit costs are 6.3 and 10.7 Pf/unit. Circulation frequency cannot however be considered as a static figure. For example, an increase in circulation frequency to 15 trips for both systems, would only mean marginal cost decreases of 0.2 Pf/unit for the glass bottle and 0.3 Pf/unit for the returnable plastic bottle.

ECOLOGICAL ASPECTS

The problems associated with the increasing amount of refuse from liquid food packaging have started discussions about the best kind of beverage containers from an environmental aspect. Because the packaging and handling of fresh milk must meet particularly stringent requirements, the study of the re-introduction of returnable packages for milk has to be considered as a test under very difficult conditions and the results discussed here refer exclusively to milk and are *not* applicable to other drinks.

Relevant criteria for estimating ecological problems are pollution in terms of amount of refuse, waste water load and emissions on the one hand, and protection of resources, for example consumption of water, raw material and energy, on the other hand. It is outside the scope of these considerations to describe all parameters of influence, not least because different scientific investigations of the same factor often come to quite different conclusions. In addition, the experts do not agree how to assess the relative importance of influencing factors when there are several conflicting factors. From available data, some facts concerning the amounts of refuse and energy consumption have been chosen to demonstrate the dependence of ecological assessments on the defined preconditions, especially circulation frequency of returnable bottles.

The amount of household refuse disposal arising from one-trip cartons depends on the proportion of refuse that is incinerated. Under average conditions in Germany, approximately 25% of refuse is burnt, so about 20.2 kg refuse per 1000 litres of milk arises from one-way cartons – see Table 12.2. This value is somewhat better than the corresponding value for the glass bottle with 11 trips. But with 12 trips this parameter demonstrates an advantage for the returnable system. This clearly shows how sensitive amount of refuse is to initial assumptions.

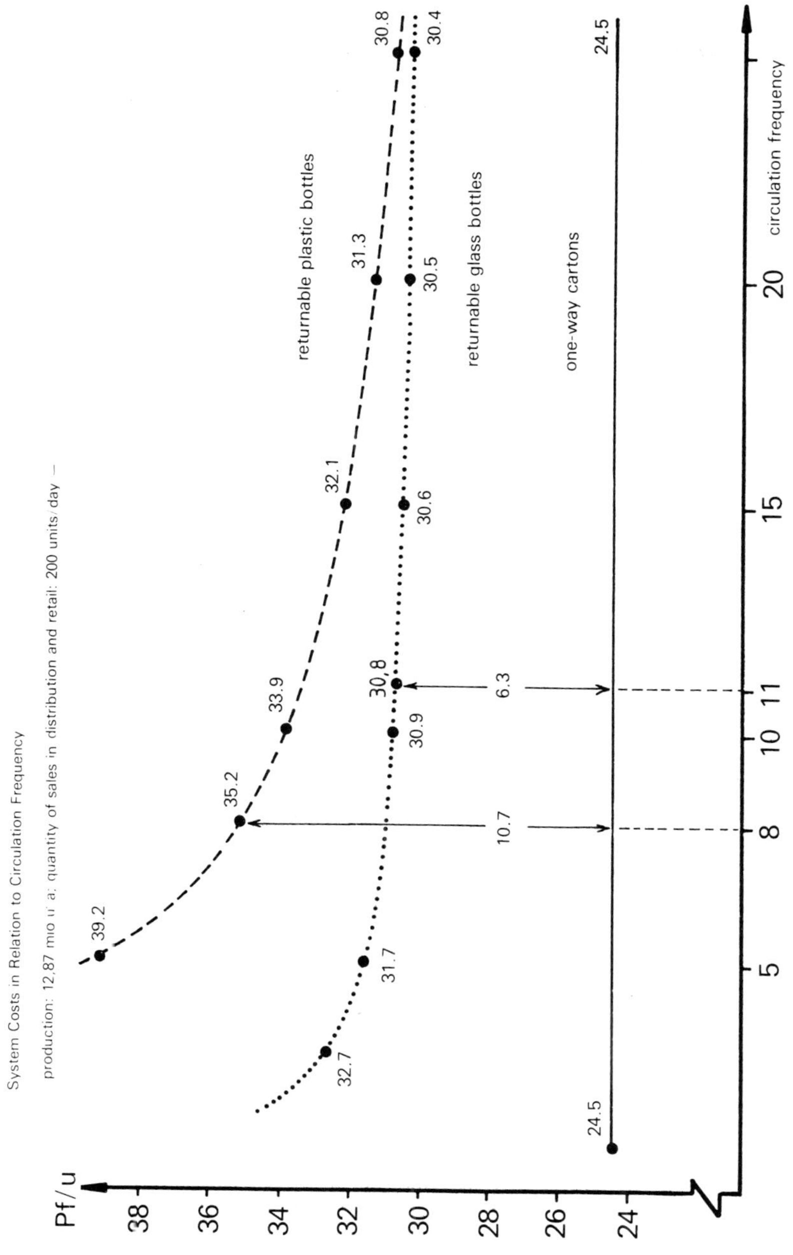

Fig. 12.6 – System costs as a function of circulation frequency for milk delivered in refillable bottles.

Table 12.2

Household refuse production in kg per 1000 litres of milk delivered in different types of packages. Source: Thome-Kosiensky (Ref. [3])

Container	% refuse incinerated	trippage	Refuse produced (kg/1000 l)
One trip carton	0	–	27.0
	25	–	20.2
	50	–	13.5
Returnable glass bottle	–	10	23.1
	–	11	21.5
	–	12	20.0
	–	15	15.4
	–	25	9.3

Energy consumption of alternative milk packaging systems is shown in Table 12.3. For the one-way carton the energy requirement depends upon whether or not the energy content of the raw material, wood, is taken into account. If the energy content of wood is excluded, energy recovery from refuse burning cannot be taken into consideration. Under such conditions the energy consumption of the one-way carton amounts to 0.46 kWh per litre of milk. On the other hand, the energy consumption of the carton may be estimated at 0.64 kWh per litre of milk, when the energy content of the raw material, wood, is included in the assessment, and it is assumed that 25% of refuse is burnt. For the glass bottle, with 11 trips, the system assessement is again dependent on the initial assumptions and under realistic conditions, none of the systems exhibits a clear advantage in terms of energy consumption.

OVERALL ASSESSMENT

According to the preliminary results from two environmental aspects of alternative packaging systems, and based on a knowledge of the present state of discussion of the other ecologically relevant factors of influence, it seems appropriate to conclude that with the circulation frequency found in the Köln market survey, there are no general environmental disadvantages for the one-trip carton. Therefore, if individual enterprises must decide whether to reintroduce milk in returnable bottles, they need consider only economic factors.

When considering economic aspects, the calculated costs must be compared with expected profits. As a first step, it is assumed that the data concerning consumer acceptance and system costs, obtained from the Köln market survey, are generally valid. A simple comparison between extra costs of the multi-trip system (about 6 Pf/unit for the glass bottle) and the corresponding extra profits (also 6 Pf/unit) shows that there are practically no differences between the alternative packaging systems from an economic viewpoint.

Since a comparison such as this is based on a number of assumptions and, since only some of the factors relevant to the decision have been considered, a risk analysis must be made, weighing all chances and potential risks in investment for a diary enterprise. Such an analysis shows that there is no guarantee of additional sales. In addition, there are considerable risks in terms of achieving the desired product quality. Moreover there is also the possibility of regional variations which could limit the applicability of the test results.

These considerations lead to the conclusion that, under current conditions, it is generally inadvisable for dairies in the Federal Republic of Germany to reintroduce the returnable packaging system for milk. Only where certain conditions reduce investment risk, for example existing plants and buildings, is it worth considering taking advantage of the particular market chances offered by returnable bottles.

Table 12.3

Energy requirements in kWh per litre for alternative milk packaging systems. Calculations cover all operations from the extraction of raw materials through to final waste disposal. Source: Sundstrom (Ref. [4])

System	Energy (kWh/litre)
One trip carton	
Total input energy:	
(a) with waste disposal	0.67[a]
(b) with incineration	0.54[a]
Total output energy	
(a) with waste disposal	0.46
(b) with incineration	0.43
Returnable glass bottles	
with trippage = 10	0.54
11	0.53
15	0.48
20	0.44
25	0.42

[a]With 75% disposal and 25% incineration, energy is 0.64 kWh/litre

REFERENCES

[1] G. F. K.-Nürnberg: Abschulßberichte zum 'Marktversuch Milchmehrwegverpackung', unveröffentlichte Manuskripte, Nürnberg, Juli, 1981.

[2] Wietbrauk, H. und Longuet, D. Systemkosten der Einweg-/Mehrwegverpackung von past. Konsummilch unter besonderer Berücksichtigung der Flaschenumlaufhäufigkeit. *Kieler Milchwirtschaftliche Forschungsberichte,* **34** (2), 199–262 (1982).

[3] Thome-Kosmiensky: Umweltauswirkungen durch Frischmilcheverpackungen – Systemvergleich. Berlin, 1981.

[4] Sundström: Milchverpackungen und Energie in der Bundersrepublik Deutschland. Malmö, 1981.

CHAPTER 13

The glass milk bottle is less healthy than cartons

W. J. M. DUYVESTIJN
Melkunie Holland b.v. De Bleek 1, Postbus 222, 3440 AE Woerden, Holland

INTRODUCTION

Recently the assertion that glass is a 'healthy' packaging material has been criticised. A number of earlier investigations have shown that glass as well as cardboard/polyethylene composite are 'healthy'. However, it appears that glass bottles by no means are so inert as is frequently claimed.

Dr. E. Wartenberg asks in his paper published recently in German specialist literature 'Is glass healthy?' [1]. The question arose when comparing the results obtained from migrational investigations on milk packages made of glass with those made of a composite of cardboard and polyethylene [2]. Although the results of these investigations clearly show that glass bottles and cartons are suitable for milk packaging because both packaging materials easily comply with the standards for food packages, they provided Dr. Wartenberg with a motive to criticise some German publications where glass bottles were recommended as milk containers, because glass allegedly is the best packaging material for toxicological and organoleptic reasons [3].

Dr. Wartenberg's publication is the first successful attack against the deadlocked opinion regarding glass bottles within the framework of the international discussions on one-way milk cartons and returnable glass bottles. His criticisms can be summarised as follows:

The two packaging materials which are in contact with the milk, i.e. glass and polyethylene, are polymers; glass is an inorganic polymer, polyethylene is an organic polymer. Soft glass, a type of glass with low chemical resistance, is used for the production of milk bottles for reasons of cost and processibility [4]. The cardboard carton is coated inside and outside with low density polyethylene. Both soft glass and polyethylene are synthetic products and do not exist in nature. They can therefore be regarded as synthetic materials.

Soft glass is made of sand, lime and soda with an addition of metal compounds. Soda is used as a softener to lower the high melting temperature and to simplify the processing of glass. Polyethylene is made of ethylene in a high-pressure process. Peroxides are needed for polymerisation and they act as catalysts, which are decomposed during production. Apart from a small quantity of anti-oxidants, neither softeners nor other additives are added.

MIGRATIONAL INVESTIGATIONS

When the total migration of impurities from glass milk bottles is compared with those from cardboard/polyethylene composite the value for the glass bottles is more than eight times higher than those for the cartons. For glass bottles the migration rate is 2.5 mg/l while for cardboard/polyethylene, the migration rate is 0.3 mg/l.

The total migration of both packaging materials were determined under the same standard conditions; 10 days at 40°C in H_2O. There will therefore be considerably more non-food constituents in milk packaged in glass bottles than in milk packaged in cartons. These findings are in flagrant contradiction to the previous, general assertion, namely that soft glass does not give off any constituents to the products packaged.

When determining the total migration after half a year (important for long life milk products), the value for cardboard/polyethylene composite of about 0.3 ppm remains constant [5], whereas that for glass rises and exceeds 30 ppm [6, 7]. The migration products from glass consists mainly of silicon dioxide and metal oxides. The migration products from polyethylene composite contain soluble, waxlike polymers (no monomers) of low molecular weight, which – as proven by animal tests – are not deleterious [8].

However, for *hard* glass, migration values are comparable to those of the cardboard/polyethylene composite.

A further remarkable difference between the two packaging materials is the share of heavy metals. Although the share of heavy metals in both packaging materials is recognised as physiologically safe, the values for soft glass are higher as shown in Table 13.1. When comparing the analyses of heavy metals in the aqueous migration products from both packaging materials it is clear that the values are below the detection limit in both cases and thus far below the limit values according to the German regulations for drinking water as shown in Table 13.2.

FLAKING AND OTHER ASPECTS

Soft glass can also give off insoluble products to the milk besides the soluble migration products. This negative phenomenon is called flaking, which means that extremely thin flakes are formed inside the glass bottle. Flaking can take place in soft glass during sterilisation when pouring a hot fluid in a bottle, but also during long storage at room temperature. In 1979, 5000 cases of liquor were withdrawn in the United States due to flaking [9–14].

As far as the organoleptic qualities are concerned, it has never been proved that glass is the best packaging material for food in general and for milk in particular. Scientific investigations made on milk containers have even proved the contrary. As milk is very sensitive to light (oxidation taste due to the influence of light) and as the one-way carton made of cardboard/polyethylene composite gives a good protection against light, the latter container should be preferred for organoleptical reasons when compared with the glass bottle.

Table 13.1

Heavy metals present in soft glass and polyethylene

Metal	Concentration in mg/kg	
	soft glass	polyethylene
Arsenic	1.8	<0.1
Lead	5.0	<1.5
Barium	85	<1.0
Chromium	7	<1.0
Cadmium	0.5	<0.15
Mercury	<0.006	<0.03

Table 13.2

Migration of heavy metals from soft glass and from polyethylene

Metal	Migration in 10^{-6} grams/litre (parts per billion)		
	from cardboard/ polyethylene composite	from soft glass	Limiting values according to German regulations for drinking water
Arsenic	<0.5	<0.5	40
Lead	5	5	40
Barium	5	5	–
Chromium	5	5	50
Cadium	1	1	6
Mercury	0.2	0.2	4

SUMMARY

On the strength of the aforementioned facts, further confirmed by Professor Scholze [15], it must be concluded that glass bottles are considerably less inert than is often claimed, and that glass bottles with regard to their toxicological and organoleptic properties certainly are not superior to one-way cartons.

REFERENCES

[1] Dr E. W. Wartenberg, Verpackungrundschau 5/1981, p. 574 t/m 577.
[2] Forschungisinstitut, Berghof gmbh, Tübingen.
[3] H. Glaseser und H. Zeiler, 'Deutsche Molkerei Zeitung' F18/1980.
[4] Jebsen Marwedel, 'Glastechnische Fabrikationsfehler', Springer Verlag, Berlin, 1980, p. 38 and H. Scholze 'Glas', 1977, p. 82.
[5] Lars Haraldson, Analytisk Kemi, University of Lund, Sweden.
[6] B. Moody, *Packaging in Glass,* Hutchinson Benham, London, 1977, p. 82.
[7] Chemistry of Food Packaging, *Advances in Chem. Ser.,* **35**, 1977, p. 19.
[8] R. Lefaux, Les Matières Plastique dans l'industrie alimentaire, 1972, p. 448 and publ. Tech. Associés, Paris.
[9] Turner, *Journal of Am. Soc.,* **18**, 1935, p. 135–141.
[10] *Referate Dienst Verpackung,* 2/1977 No. 1 p. 100:0347 and No. 3 p. 121:1348.
[11] Bacon, *Glass Ind.* (Aug-Oct 1968), and *Chem. Durability of Silicate Glass.*
[12] A. Peters, *Glastechn Berichte,* **16**, 1973, 161.
[13] *Food Chem. News,* **20** (45), p. 2, 1979.
[14] 'Flüssiges Obst' 9a/80.
[15] Prof. Dr H. Scholze, *Chemische Beständigkeit von Glas, Grundlage seines Einsatzes aus Verpackung.* Verpackungs-Rundscha 8/1981 p. 880–883.

The Author

Ir W. J. M. Duyvestijn graduated from the University of Agriculture at Wageningen in 1968 and is now Head of Packaging Development at Melkunie Holland in Woerden.

CHAPTER 14

Evaluation of the approach to the problems of milk packaging and the environment in the Netherlands

D. VAN DOORN AND H. MIDDELVELDT
Produktschap voor Zuivel, Rijswijk, Holland

INTRODUCTION

Many people are being confronted with environmental problems these days, whether they like it or not and the dairy industry is no exception. Changes in consumption and sales patterns occurred in the food sector in the 1960s and 1970s, a development characterised by the sale of more and more packaged products which has greatly increased the volume of household refuse.

Discussion between government and industry on controlling the rising amount of waste began in an emotional climate which entailed the risk of measures failing to do justice to all the relevant factors and thus being incapable of providing a solution to the problem. The Dairy Produce Commodity Board, by adopting a fundamental approach, has succeeded as an intermediary in leading the discussion between government and industry into more businesslike channels. It could do so because of the knowledge available to its secretariat about milk and milk packaging.

The primary job was to start a dialogue between government and the industry, in which the problem could be approached jointly on the basis of an objective analysis. The approach comprised three main elements:

(a) The Commodity Board's executive committee set up a steering group named 'Milk Packaging and Environment' and comprising representatives of government and all sections of the industry. The aim of the group was to guide and stimulate discussion of this subject and to make a recommendation to the Commodity Board's executive committee which advises the government on policy lines to be adopted.

(b) It was decided to prepare a report on 'Milk Packaging and Environment' which would probe the relevant factors.
(c) An agreement was made between industry and government to the effect that consultation will take place before any new container is actually introduced.

The report was intended to form the basis of the discussions but since little time was available for its preparation, it was thought necessary to probe a number of factors more deeply. For this reason the Commodity Board's executive committee concluded at the end of 1981 that, in response to urging by government and industry, the following aspects should be studied in more detail:

(i) Cost comparisons between one-trip cartons and returnable glass milk bottles were not supported by sufficient evidence. The Commodity Board had reliable figures up to the retail trade channel itself and was able to give notional values which subsequently proved to be reasonable. Data for the milk retail trade were however lacking.
(ii) Before experimenting with returnable plastic milk bottles, a brief cost analysis should be made between this container, cartons and glass bottles. It was not thought meaningful to start an experiment with an alternative in the absence of such information, when it was uncertain whether the alternative could be an improvement. In addition, the plan for the experiment would cause further study as the Ministry of Public Health and Environment felt it could be improved on.
(iii) Too little attention had been paid in the past to social costs measured in terms of raw material use and waste disposal. This aspect should be examined.

As the Commodity Board had no expertise in the area of waste-disposal technology, the Ministry of Public Health and Environment agreed to provide more information to enable the steering group to assess the social cost. Meanwhile, the secretariat of the Dairy Produce Commodity Board completed research on the first two points and duly advised the steering group. These points are discussed later. The following meeting of the Commodity Board's executive was also scheduled to discuss these points after which there were to be discussions between the State Secretary of Public Health and Hygiene and the Commodity Board's executive committee at the beginning of July 1982.

The approach favoured by the Commodity Board's executive committee was appreciated by both government and industry. The responsible Minister has intimated that she considers this the right way, which should be continued. Only in this way is it possible to approach a solution step by step. Her predecessors had believed that the problem could be solved by stimulating the use of returnable packaging, whether voluntary of mandatory. Today, it is recognised that the problem should be approached in a broader way. The present Minister shares this view but still attaches some importance to returnable packages.

So it is expected that she will suggest at forthcoming talks with the executive committee that returnable packaging of school milk be substituted for the present one-trip packaging. One of the arguments is the educational aspect. It is still too early in the discussion to involve this last point in the evaluation. The steering group and other advisory bodies will first have to discuss the idea on the basis of a note giving all the advantages and disadvantages.

At this stage, it is worth making some observations on a proposal for an EEC directive on beverage packaging, which the EEC Commission has put to the Council of Ministers. It is well known that it was originally the Commission's intention to introduce specific mandatory regulations, such as the prohibition of some sealing systems or the introduction of a mandatory deposit for drink containers. The proposals now indicate that it is left to member states to take specific measures to curb the rising flow of waste.

However, this would create an opportunity to undermine one of the principles of the Rome Treaty, namely, free trade among the member countries. On the other hand, recent judgements of the European Court of Justice seem to lead to the conclusion that the possibilities for issuing regulations to member states are quite limited in this area. Truly a viscious circle. Hence we would like to recommend that the EEC deal with this matter in the way that the Dutch dairy sector has dealt with it. It may not be the shortest road, but it is certainly a practicable one.

MILK PACKAGING

The Dairy Produce Commodity Board carried out a study during 1980 and 1981 on the many aspects of the packaging of liquid-milk and liquid-milk products. The results of this study was laid down in a report entitled 'Milk Packaging and Environment', which was published in May, 1981. Some of the subjects discussed in the report have been the focus of interest in the Netherlands this past year. The Dairy Produce Commodity Board collected data on some points and made a closer study of some others. The following is a report on developments in the Netherlands this last year with regard to Milk Packaging and Environment.

THE DAIRY PRODUCE COMMODITY BOARD

A Commodity Board is an institution, established by law, of and for a sector of trade or industry. These sectors – as the name 'produktschap' indicates – concern a product or a group of products, as for instance 'dairy products', 'cattle and meat' or 'vegetables and fruit'. Such boards have been formed especially in the agricultural sector. They are public bodies and the executive committees are competent to lay down regulations. These committees are formed of people belonging to the sectors concerned. The Dairy Produce Commodity Board was instituted in 1956. It is made up of people who in their daily work are concerned with milk and milk products, such as butter, cheese, condensed milk, milk powder and liquid milk. In this board, dairy farmers, manufacturers of dairy products and wholesale traders and retailers in milk and dairly products, as well as their employees (manual workers, administrative and technical personnel) work together to look after their common interests. They have an interest in how the board performs its duties because it directly affects the development of the industries in which they work.

Table 14.1 shows the sales of liquid milk and liquid milk products in 1981 and it has been computed that these home-produced milk packages produce 48,500 tonnes of household waste per annum. In addition, imported milk gives rise to a further 4,000 tonnes of household waste. These 52,500 tonnes of waste represent about 1.3% of the total volume of Dutch domestic waste, which stood at 4,000,000 tonnes in 1981. About 8% of domestic waste consists of disposed beverage packaging and, of the total category of beverage packaging, milk packaging accounts for around 20%.

Table 14.1

Sales of liquid milk and liquid milk products by the Dutch Dairy industry in 1981

Product	Sale in tonne
Liquid milk	942000
Cream	36000
Sour milk products	369000
Porridge & custard	215000
Chocolate milk	67000
Total	1629000

Table 14.2 shows the type of packaging used for milk and milk products. As can be seen, the share of returnable packaging is declining slowly. In the Netherlands about one quarter of the liquid milk is sold by the specialised dairy retail trade. This outlet uses predominantly returnable packages. Table 14.3 shows the proportion of dairy products sold in different outlets.

DETAILED SURVEY OF RETAIL TRADE COSTS CONNECTED WITH MILK PACKAGING

Chapter VI of the 'Milk Packaging and Environment' report published in May, 1981, discusses cost differentials between pasteurised milk in 1-litre cartons and

Table 14.2

Proportions of milk and milk products marketed by dairy factories during 1970–1981 in different types of packaging

Year	Percentage of total marketed			
	Unpackaged	In returnable packages	In one-trip packages of all types	In cartons
1970	4	71	25	14
1973	1	52	46	34
1976	0	43	56	42
1979	0	35	65	53
1980	0	33	67	55
1981	0	31	69	57

Table 14.3

Proportion of milk and dairy products sold for domestic use from different types of outlet

Outlet	Percentage of total sales	
	1980	1981
Delivery trade	14.3	12.6
Mobile shops	14.9	12.5
Shops	70.8	74.9
Totals	100.0	100.0

in 1-litre returnable glass bottles. Regrettably, cost differentials given for the food trade channel were based mainly on a study carried out for the Central Food Trade Bureau (C.B.L.). The returnable bottle was 11 cents more expensive than the one-way carton. As the Dairy Produce Commodity Board lacked time to make its own study, it agreed to use these data. Chapter VI also noted that reliable figures on cost differentials between the various kinds of milk packages used in the milk trade channel were not available.

Today more detailed information is available on cost differentials in both the food and milk trade channels. In studying costs in the food trade channel the Commodity Board received cooperation from both the C.L.B. and various food retail businesses. The cost study for the specific milk retail was made in close cooperation with the Industrial Board for the retail Trade in Milk and Dairy Produce.

Table 14.4

Extra cost of returnable glass bottles versus cartons for milk sales in the food retail sector

Operation	Extra cost of glass bottles	Range
1. Intake and sorting	5.50	4.9–7.2
2. Cash desk registration and refund of deposits	0.60	0.35–1.45
3. In-store handling; loading and unloading trucks	0.20	0–0.3
4. Use of space and other accomodation costs	2.00	0.9–3.0
5. Breakage and spillage	0.45	–
6. Refrigerated display case plus refrigeration	1.50	1.3–1.7
Total	10.25	

Table 14.5

Extra costs of milk packaging in 1-litre returnable glass bottles compared with 1-litre cartons

Cost area	Extra cost of glass bottles (in guilder cents per litre)
1. Labour costs of glass system	4.54
2. Storage of finished products	0.51
3. Vehicle costs	1.80
4. Accommodation costs	0.39
5. Storage cost of empties	0.09
6. Breakage	0.04
Total	7.37

Notes for Table 14.5

1. Labour costs relate to intake of bottles, refund of deposits, sorting and accounting.
2. Relates to storage cost of fresh and non-perishable milk and milk products and includes depreciation, maintenance and interest on refrigerated and other stores.
3. Includes depreciation, maintenance, insurance and tax on fuel.
4. Relates to accommodation costs of fresh and non-perishable milk and milk products such as the cost of electricity for refrigeration, refrigeration of mobile shops, lighting and cleaning costs.
5. Calculations are based on the price per square metre in stores or public taxes on outside storage.

(a) Extra cost of returnable packages (1-litre glass bottles) compared with those of one-trip packages (1-litre cartons) in the food trade channel.

Table 14.4 reproduces a survey complied by the Commodity Board's Secretariat on the basis of cost figures by various food businesses. The cost figures given are averages. Column 2 shows the spread of costs to indicate the wide differences between the figures obtained.

(b) Extra cost of returnable packages (1-litre glass bottles) versus one-trip packages (1-litre cartons) in the milk trade channel.

Table 14.5 shows the extra costs of returnable over non-returnables and has been compiled in close consultation with the milk retail trade. The cost study covers both retailers serving customers by mobile shops and retailers using traditional vehicles, with about 20 retailers of each type. The final figures have been compiled as weighted averages.

To approach cost differentials between glass bottles and cartons for the kinds of costs stated under 2 to 5 in Table 14.5, differences in the use of space have been taken as a basis. These cost differences have been computed from the integral cost figures assuming that 1.6 times as much space is needed to store a unit of bottled milk as for the same quantity packed in cartons and assuming that the size of the kinds of costs in question are proportional to the use of space.

(c) Comparison of cost differentials in food trade channel and door-to-door milk delivery.

The costs of Tables 14.4 and 14.5 have been reduced to three categories to allow comparison, and these are shown in Table 14.6.

Table 14.6

Extra cost of milk packaging in 1-litre returnable glass bottles compared with packaging in 1-litre cartons; comparison of shop sales with door-to-door sales

Cost area	Extra cost in guilder cents per litre	
	Shop retailing	Door-to-door retailing
Labour costs	6.30	4.54
Storage, refrigeration, selling space	3.50	2.79
Breakage/spillage	0.45	0.04
Totals	10.25	7.37

POLYCARBONATE MILK BOTTLE

Discussions of milk packaging, regularly mention two factors which have caused the steady decline in the share of the returnable glass bottle; these are its high weight (approx. 630 gramms) and the risk of breakage.

It may be possible to reduce the weight of the present standardised milk bottle somewhat, but a plastic bottle and a carton offer considerable advantages in terms of weight. At the same time, there might be a higher breakage rate with a lighter glass bottle.

The Ministry of Public Health and Environment has observed that the above aspects are scarcely if at all applicable to returnable plastic (polycarbonate) bottles. The Ministry raised the question of whether introducing the polycarbonate bottle might not be a means of at least stabilising and possibly enlarging the market share held by returnable packaging. Representatives of the liquid-milk industry in the Netherlands, influenced to some extent by publications of trials in the United States showing that returnable plastic bottles used for milk entail a greater risk of problems than do glass bottles, were doubtful whether a polycarbonate bottle could provide an attractive alternative form of milk packaging in this country. It was felt that, before definite opinions could be given, a study should be made by the Netherlands Dairy Research Institute (NIZO) to establish with certainty that the polycarbonate bottle, when used as a returnable milk bottle, would not present any technical or qualitative problems. Having learned of the results of an experiment with returnable glass and polycarbonate milk bottles in West Germany, and after some literature study, NIZO proposed experiments with polycarbonate milk bottles.

The purpose of the experiment was to find out whether returnable polycarbonate (pc) bottles qualify as a substitute for current packages of milk and milk products. The requirements demanded of the bottles are that (a) they should have the same advantages as cartons, namely lightweight, square, airtight if possible and not imparting a taste to the packaged product, and (b) they must exhibit clear advantages over glass bottles.

The products used in the tests on pc bottles should conform as closely as possible with normal practice. To obtain a good impression of the average useful life and quality of used bottles, they have to be washed, filled, handled, stored and emptied about 100 times. This way of testing will yield reliable information for correctly comparing the cost of the use of pc bottles with that of glass bottles. To keep the cost of the project down, bottles and milk are organoleptically and microbiologically tested only on the first, the 25th, the 50th, the 75th and the 100th trip.

The bottles used will be 1-litre polycarbonate bottles of square cross-section; 2,000 bottles will be employed including 50 brown-coloured and 150 colourless, marked NIZO in blue lettering.

Two types of milk will be used in the tests:

(a) Non-homogenised milk as normally produced in the dairy factory.

(b) Homogenised, partly-skimmed milk for trip Nos. 1, 25, 50, 75 and 100, because possible abnormalities of smell or taste are easier to trace in such milk.

Bottles are treated as follows on each trip:

1. The 2,000 bottles are washed and filled on a modified production line in a dairy factory.
2. The 50 brown and 150 marked colourless bottles (referred to later as 'the 200') are hand-packed in crates and stored in the factory at a temperature of 5 to 7°C for 25 hours. The remaining 1800 bottles are emptied and washed and remain in the factory until the next trip.
3. After cold storage 'the 200' are taken to NIZO in their crates and exposed to daylight for 1 to 2 hours ('on the retailer's doorstep').
4. The bottles are then taken from their crates and kept under artificial light at 5°C ('refrigerated display case in the shop') for five days.
5. (a) On trip Nos. 1, 25, 50, 75 and 100, the milk is tested organoleptically and microbiologically. The bottles are tested microbiologically for cleanliness and their appearance is inspected (scratches, turning grey, etc). Migration analyses will be carried out on these trips, using media which are representative for neutral, acid and fatty products. Finally the bottles will be emptied and washed. (b) On all other trips, the bottles will be merely emptied and washed.
6. The empty bottles are taken in crates to the production plant to be used on the next trip together with the 1800 other bottles which have remained in the factory.

The executive committee of the Dairy Produce Commodity Board has decided in principle to have the experiment carried out but, in consultation with the Ministry of Public Health and Environment, it was decided to postpone the start of the experiment pending a further cost analysis of the pc bottles.

COST OF THE RETURNABLE SQUARE PLASTIC BOTTLE

During discussions of the proposed experiment with the returnable plastic bottle, it was felt desirable to estimate the cost of using this container before the experiment started. The problem is that no data for the proposed container are currently available. Calculations have therefore been made as shown in Table 14.7 using the known cost of glass bottles for labour costs and carton data for use of space.

Table 14.7 showing differentials for different cost elements in the dairy factory, proves that the cost of 100 litres of milk in square polycarbonate returnable bottles is fl. 2.35 less than when packed in cartons. Differentials between pc and glass are minimal in the dairy factory. It is noted that this survey is based on the cost as stated in the 'Milk Packaging and Environmental' report of the Dairy Commodity Board. The following points should be noted in relation to Table 14.7.

Table 14.7

Production costs (1981) in the diary factory of 100 litres of milk in 1-litre returnable glass bottles, 1-litre carton packages and 1-litre square returnable polycarbonate (pc) bottles

Cost area	Production costs (fl)			Cost difference pc versus carton	
	glass	carton	pc bottle	pc bottle dearer (fl)	pc bottle cheaper (fl)
1. Depreciation/ interest	4.90	3.70	4.90	1.20	–
2. Wages	13.20	11.00	13.00	2.00	–
3. Cartons	–	8.00	–	–	8.00
4. Bottles	1.00	–	1.50	1.50	–
5. Caps	0.40	–	0.40	0.40	–
6. Crates	0.45	0.30	0.30	–	–
7. Electricity	0.40	0.30	0.40	0.10	–
8. Oil/gas fuels	0.50	0.30	0.50	0.20	–
9. Water	0.40	0.25	0.40	0.15	–
10. Repairs and maintenance	1.70	1.45	2.70	0.25	–
11. Other costs	3.95	4.10	3.95	–	0.15
Totals	26.90	29.40	28.05	5.80	8.15

(a) 1-litre carton and glass containers
Calculation of the cost of producing and packing is based on actual costs as fixed annually by the Commodity Board.

(b) 1-litre returnable pc bottle
As this container is not used in the Netherlands, the cost of production and packing has been estimated as closely as possible.

It has been assumed that no major investment is needed, because the existing machines for glass containers can be used for handling, filling, sealing and washing. It has also been assumed that the purchase price of a pc bottle is twice that of a glass bottle and that a pc bottle will make 60 trips (glass bottle 40 trips).

(c) Transport costs
The 'Milk Packaging and Environment' report puts transport costs (i.e. of transport from dairy factory to shop and from shop to dairy factory of returnables) at:
fl. 8.50 per 100 litres for 1-litre returnable bottle;
fl. 5.50 per 100 litres for 1-litre cartons.
Transport costs of the 1-litre returnable pc bottle have been estimated at fl. 6.00 per 100 litres.

(d) Cost to the food trade channel
The extra cost of returnable bottles to the food trade channel are as shown in Table 14.8.

Table 14.8

The extra costs of 100 litres of milk in 1-litre glass returnable bottles compared with 1-litre polycarbonate (pc) returnable bottles for sale through food shops

Cost area	Costs/fl	
	glass bottle	pc bottle
1. Intake and sorting	5.50	5.50
2. Cash-desk registration and deposit refund	0.60	0.60
3. In-store handling and loading/unloading trucks	0.20	0.20
4. Store accommodation	2.00	2.00
5. Breakage and spillage	0.45	0.10
6. Refrigerated display plus other other refrigeration	1.50	–
Totals	10.25	8.40

Table 14.9 summarises the cost differentials for the different types of containers. This table shows that the total cost of milk in pc containers is approximately fl. 6.05 higher per 100 litres than that of milk in cartons. Compared with glass, the pc bottle is fl. 4.70 cheaper. The above cost differentials will be one of the factors on which it will be decided shortly to carry out the experiment with the polycarbonate bottle. The view of the Ministry of Public Health and Environment will play a major role in this decision. If it is decided to go ahead with the experiment, it remains to be seen how the consumer will react. Under present plans, the experiment will last 28 months and cost about fl. 400,000.

Table 14.9

Total cost differentials between 100 litres of liquid milk packed in cartons, returnable glass bottles and returnable polycarbonate (pc) bottles

Cost contribution	Cost/fl		
	carton	pc bottle	glass bottle
Supply costs (factory to plant)	1.93	1.93	1.93
General factory share	7.80	7.80	7.80
Production and packing	21.60	19.25	19.10
Transport costs	5.50	6.00	8.50
Cost total excluding retailer	36.83	34.98	37.33
Extra cost at retailing	–	7.90	10.25
Total costs	36.83	42.88	47.58

GLASS RECYCLING

Bottle bank campaign

An intensive three month campaign is being waged in the Netherlands as from March 31, 1982, to promote greater use of bottle banks for glass containers. The campaign is being undertaken by government and business jointly under the slogan 'Give glass a chance again. Into the bottle bank'. In support of the campaign, the Ministry of Public Health and Environment has undertaken:

(i) To invite municipalities which have not yet set up bottle banks, to start glass collection now.

(ii) To urge municipalities which are collecting already, to improve collection by setting up sufficient numbers of bottle banks at suitable places and making collection of the containers more efficient.

(iii) To invite municipalities to lend support to the campaign by local publicity through information and local media. To this end, standard campaign advertisements and big stickers on refuse trucks have been designed. In addition, a 30-second spot is being screened on television at least 18 times during the campaign and the government is also promoting the bottle bank campaign via the radio and daily newspapers.

In cooperation with the Netherlands Local Government Association and the Netherlands Association of Refuse Collection Department Directors, the action is being given great attention in trade journals, and those conerned are urged to take measures to stimulate the citizens. Businesses take part in the action through organised activities. All shops, about 10,000, selling products in one-trip glass containers are given free material designed to draw shoppers' attention to the bottle bank campaign. In this way nearly all Dutch people are regularly confronted with the campaign. To support shopkeepers' efforts, full-page advertisements are inserted in the chief national daily newspapers in addition to the screening of television spots, and the Dutch media (newspapers, radio, television and trade journals) were extensively informed before the campaign was launched.

Trade and industry (glass makers, fillers of one-trip glass bottles, food stores, bottle bank operators) are raising fl. 500,000 to finance these activities. The value of the governmental activities is estimated at fl. 250,000.

Government and business have drawn up the following 10 reasons for collecting one-trip glass containers.

(a) Reduction of the waste mountain, because if glass is kept apart from other waste and used to make new glass, a much lower volume of refuse needs to be collected.
(b) A saving of over 35% on energy consumption for glass melting and raw materials production, if new glass is made entirely of waste glass.
(c) Nature conservation because less silica and lime from natural sources is needed if waste glass is recycled.
(d) Less environmental pollution because soda, production of which pollutes the environment, is a raw material for glass production which can be dispensed with if old glass is re-used.
(e) Considerable savings on the cost of collecting and processing household refuse if glass is kept apart from other garbage.
(f) Less wear on furnaces if waste offered for combustion contains no glass.
(g) Better compost; if household refuse is turned into compost the quality is better as it contains less glass.
(h) People putting bags of refuse on their doorstep and dustmen are less likely to be injured if bags contain no glass.
(i) Glass retains its quality even if melted over and over again.
(j) This form of 'separation of source' is recommneded and successfully carried into effect also at an international level as in the EEC.

Some 100 million kg of glass were collected in the Netherlands in 1981, which amounts to a response of about 30% in municipalities which have set up bottle banks. Bottle banks have been set up in about 80% of all Dutch municipalities to date and the policy is to have banks in every municapality. The glass collection system comprises on-the-spot emptying of the bottle banks once a week. More than one bank is placed if deposits exceed the capacity of one container (which does happen here and there).

The normal number of bottle banks is based on a density of one bank per 2,000 population. The glass collected in this way goes to processing plants where the glass is first crushed to suitable size. Metal is removed by means of a magnet, and other foreign matter such as corks, plastic caps etc. are winnowed out by air blasts. The cleaned cullet, as the broken glass is called, goes back to the glass-making industry where new glass of green or brown colour is made from it. Cullet is not suitable for making white flint glass. However, demand for green glass still exceeds suppliers of cullet, so this raises no problems.

Should supplies of old glass ever exceed demand for green glass, then it will become necessary to sort cullet by colour. This could be done either by separating glass for colour at source, or separating glass for colour at the glass-making plant. The recycling industry feels the latter is feasible and development of suitable machinery is under way. Table 14.10 summarises the results of glass collections in the Netherlands.

Table 14.10

Results of glass recycling in the Netherlands

Year	1979	1980	1981
Number of towns involved	100	300	500
Total number of bottle banks	800	4000	6000
Glass collected (tonne)	25	75	100

PAPER RECYCLING

A large-scale experiment with a household refuse separator was carried out in the Netherlands lasy year, using the Fläkt system. Although a report on the experiment is not yet ready, it is known that problems have been greater than foreseen. Principal amongst these was that the paper sorted out by the separator proved hard to sell. It should be borne in mind that in the Netherlands, more than 50% of the paper is separated at source. Many schools, churches and clubs collect old paper and sell it to the paper trade to raise funds. The results is that the paper content of household garbage is small and of inferior quality. Moreover, old-paper collection assures paper makers of large quantities of good-quality wastepaper, so that they have little interest in the more polluted paper from the separator.

The separator experiment is however to be followed up in two ways:

(a) The city of Groningen is engaged in an experiment of having households separate their refuse into two components: a dry component (paper, plastic, etc.) put in a refuse bag of one colour, and a wet component (kitchen and garden refuse etc.) put in a bag of different colour. This will make it possible to obtain less polluted paper from household garbage.

(b) A so-called old paper recycling plant can be set up in combination with the separator. It is technically feasible to recycle old paper to such an extent that it can replace the cellulose pulp raw material in paper making, so that it no longer needs to compete with collected old paper. The refining is being studied chiefly for its economic aspects.

SUMMARY

This report then has briefly considered the way in which problems in relation to milk packaging and environment have been dealt with in the the diary sector in the recent past in the Netherlands.

Index

A

Agence National pour la recuperation et l'elimination des déchets (France) 85
Agence pour les economies d'energie (France) 84
Agency for Environmental Protection (Denmark) 51
Alberta 50
ALKO (Finland) 74
alternative processing routes 14
aluminium 48, 49, 58, 110
aqvavit 51
arsenic 155
Australia 47
Australian Environmental Council 48

B

bans 47, 49, 50, 51, 60, 80
Barex 77
barium 155
beer 54, 55, 56, 62, 63, 68, 75, 86, 91, 94, 105, 127
beverage containers 26–45
board 85, 86, 99
Bottle Bank 57, 80, 168
British Columbia 50

C

cadmium 155
Canada 49
can bank 57
capital investment 41
capital obsolescence 41
cartons 63, 89, 140, 153, 161
chromium 155
conservation policy 18
consumer 17, 22, 23, 48, 142
consumer convenience 39
consumer convenience premium 31
consumer preference 31
consumer sovereignty 32, 37
costs 163

D

Dairy Produce Commodity Board (Netherlands) 157
Delegation aux economies de matières premières (France) 84
Denmark 51, 60–64
Department of Consumer and Corporate Affairs (Canada) 49
deposits 27, 47, 48, 49, 51, 55, 56, 60, 80, 87, 121, 125, 136
deposit company 56, 117
disinvestment costs 85
Düsseldorf 54

E

economic impacts 39
electricity production efficiency 19
employment 85
energy 38, 52, 75, 84, 85, 86, 90, 100, 103, 104, 114, 127, 151, 169
energy index 107
environmental impact 36, 38
equal shelf space 49, 50
external costs 28, 29, 38
external diseconomy 130, 132
external impacts 33
externalities 31, 32
extra social valuation 37

F

Finland 51, 72–78
flaking of glass 154
Foundation for Packaging and the Environment (Holland) 79
France 26, 46, 84–109
free market mechanism 29
fruit juice 52, 55, 63, 64, 86, 91, 99
fuels 20
function oriented systems 15

G

Germany 53, 140–152, 155

glass 48, 49, 53, 54, 56, 63, 70, 82, 85, 86, 90, 107, 132, 140, 153, 168
glass containers – lightweighting 21, 87
Glass Manufacturers Federation (UK) 57
gross energy requirement 20

H

Haga agreements 126, 128

I

internal benefits 33
internal costs 29, 33, 40
investment costs 85
Iron and Metal Industries Union (Norway) 69

K

Köln 140

L

labour 40, 90
lead 155
legislation 23
levies 125, 126, 127, 128, 129, 133
lightweighting glass containers 21
litter 27, 113, 114, 127, 129, 133, 137
Litter Control Committee (Australia) 48
litter tax 27

M

mandatory deposits 47, 80
Manitoba 50
marginal impact 132
marginal internal cost 30
marginal private valuation 30
mercury 155
milk 52, 54, 64, 86, 91, 93, 140, 153, 157
mineral water 56, 68, 86, 91
minimum cost conditions 32

N

NATO 71
Netherlands 54, 79–83, 153–171
non-refillable containers 49, 80
non-returnable containers 22, 48, 54, 68, 72, 110, 125, 127, 161
Nordic Council 74
North Brabant 55
Norway 55, 65–71

O

OECD 41, 46, 52, 58, 59, 71
Ontario 50, 51
Oregon 11, 27

P

Packforsk (Sweden) 77
paper 48, 65, 80, 85, 125, 132, 170
PET 51, 53, 54, 58, 70, 77
physical units – use of 12
PIRA (UK) 77
plastic bags 65
PLM 110
policy evaluation 38
policy options 26
pollution 36, 114
polycarbonate bottles 165
price controls 39
private costs 30
product charges 47
product oriented systems 15
PVC 53, 89, 90, 107

R

raw materials 20, 84, 90, 114
recycling 23, 27, 47, 48, 54, 55, 57, 75, 80, 87, 90, 107, 110, 114, 131, 168, 170
recycling rate 49, 111
refillable containers 49, 51, 82
repurchase value 111
retrieval rate 111
returnable containers 22, 27, 54, 55, 57, 72, 84, 89, 140, 161
reuse 48, 60, 87

S

Saskatchewan 50
shandy 91, 95
social optimum 31, 32
social opportunity cost 36, 37
soft drinks 54, 55, 56, 62, 97, 105, 127
South Australia 48
systems – resource 13
systems – specified by output 13
systems – specified by function 13, 14
system costs 144
spirits 54, 55, 98, 127
standardisation 51
steel 48, 53, 110
Sweden 56, 110–139
Switzerland 56

T

taxes 27, 47, 55, 56, 63, 64, 67, 75, 80, 131
tinplate 53, 54, 125, 132
traffic congestion 37
trippage 22, 27, 46, 55, 107, 142

U

UNIDO 79
United Kingdom 57
units 12

V

Varberg 115
Vermont 28

W

waste 36, 79, 80, 84, 100, 114, 136, 138, 148
waste disposal 27, 87, 169
waste management 47
waste separation 82
wine 54, 55, 56, 63, 64, 86, 91, 127